Studies in Systems, Decision and Control

Volume 666

The series "Studies in Systems, Decision and Control" (SSDC) covers both new developments and advances, as well as the state of the art, in the various areas of broadly perceived systems, decision making and control–quickly, up to date and with a high quality. The intent is to cover the theory, applications, and perspectives on the state of the art and future developments relevant to systems, decision making, control, complex processes and related areas, as embedded in the fields of engineering, computer science, physics, economics, social and life sciences, as well as the paradigms and methodologies behind them. The series contains monographs, textbooks, lecture notes and edited volumes in systems, decision making and control spanning the areas of Cyber-Physical Systems, Autonomous Systems, Sensor Networks, Control Systems, Energy Systems, Automotive Systems, Biological Systems, Vehicular Networking and Connected Vehicles, Aerospace Systems, Automation, Manufacturing, Smart Grids, Nonlinear Systems, Power Systems, Robotics, Social Systems, Economic Systems and other. Of particular value to both the contributors and the readership are the short publication timeframe and the world-wide distribution and exposure which enable both a wide and rapid dissemination of research output.

Indexed by SCOPUS, DBLP, WTI Frankfurt eG, zbMATH, SCImago.

All books published in the series are submitted for consideration in Web of Science.

Peter S. Sapaty

Distributed Management with Spatial Grasp Model

Springer

Peter S. Sapaty
Institute of Mathematical Machines
and Systems
National Academy of Sciences of Ukraine
Kyiv, Ukraine

ISSN 2198-4182 ISSN 2198-4190 (electronic)
Studies in Systems, Decision and Control
ISBN 978-3-032-19377-3 ISBN 978-3-032-19378-0 (eBook)
https://doi.org/10.1007/978-3-032-19378-0

This Springer imprint is published by the registered company Springer Nature Switzerland AG
The registered company address is: Gewerbestrasse 11, 6330 Cham, Switzerland

With gratitude to Lilia, my wife for more than half a century and schoolmate before, for inspiration, encouragement, and support of this and previous books as psychologist, doctor, and talented cook

Preface

The word "spatial" fundamentally relates to human existence and activity in any terrestrial and celestial spaces. It refers to the location, arrangement, and distribution of things, including human activities, people, and resources, across the Earth's surface and within physical environments, making it a fundamental concept in geography and other fields like sociology and urban planning. Human activities, from daily journeys to the migration of populations, inherently occur within a spatial context, influencing and being influenced by physical environments and leading to distinct spatial patterns and relationships that are studied to understand human behavior, resource use, and societal development.

There is definitely a need for unified philosophical, psychological, and technological approach for investigation and management of any spatial organizations and systems. We have chosen for this the developed and patented Spatial Grasp Technology (SGT) and its Spatial Grasp Language (SGL) based on spatial coverage and management of distributed dynamic systems, which proved efficient for solving problems in physical, virtual and united worlds, as in the previous books. SGT historically evolved from the WAVE approach, which allowed us more than half a century ago to practically implement citywide heterogeneous computer network and solve problems on it in mobile way, before the internet. This approach allowed us to penetrate through software and hardware of different systems and also spatially manage them.

After reviewing numerous publications on spatial ideas and systems and describing basics of SGT and SGL, the book will be showing solutions of different management and security problems in both terrestrial and celestial spaces confirming suitability and efficiency of the approach. It will compare Spatial Grasp philosophy and model with the concept of algorithm and its modifications, showing universality of the former for understanding and management of any spatial and distributed systems, with the latter just representing tools for concrete applications. The book will show resemblance of space-covering SGL scenarios to massive spatial phenomena like virus, smog, radiation, rain, smell, light, and others, which are far more general and global than algorithmic sequences. The dynamic self-organizing spatial features of SGT may also relate to non-algorithmic ideas currently growing popular especially in relation to living organisms and societies where traditional algorithmic models are becoming insufficient.

Kyiv, Ukraine
September 2025

Peter S. Sapaty

Acknowledgments

With special thanks to good colleagues, friends, and their organizations for the lasting support and active discussions of the ideas of this book:

John Page, University of New South Wales, Australia.

Steve Lambacher, Aoyama Gakuin University, Japan.

Kohei Arai, Saga University, Saga Prefecture, Japan.

Thomas Ditzinger, Springer books, Germany.

Gabriella Williams, Taylor & Francis books, UK.

Melian Lee, International Relations and Diplomacy journal, David Publishing, USA.

Judith Reveal, **Victor Welzant**, and **George S. Everly**, International Journal of Crisis, Stress, and Human Resilience, International Critical Incident Stress Foundation, USA.

Siva Shiny, Project Editor, UK.

Svitlana Tymchyk, **Nataliia Karevina**, and **Yevheniia Petruk**, Journal of Mathematical Machines and Systems, National Academy of Sciences of Ukraine.

Nadya Aronova, **Nikolay Ievlev**, Institute of Mathematical Machines and Systems, National Academy of Sciences of Ukraine.

Contents

About the Author

Dr. Peter S. Sapaty Chief Research Scientist, Ukrainian Academy of Sciences, is with networked systems for more than five decades. Except Ukraine, he worked in Slovak Republic, Germany, the UK, Canada, and Japan as Group Leader, Alexander von Humboldt Researcher and Invited or Visiting Professor and chaired SIG on Mobile Cooperative Technologies within Distributed Interactive Simulation project in the USA. He developed distributed management and control technology resulted in European patent and about 300 publications including ten Wiley, Springer, Emerald, and Taylor & Francis books. He is in Marquis Who's Who in the World and Cambridge Outstanding Intellectuals of the twenty-first Century, also with boards of six international scientific journals.

Chapter 1
Introduction

Abstract The word "spatial" fundamentally relates to human existence and activity in any terrestrial and celestial spaces. The aim of this book is to describe universal high-level model and technology for dealing with spatial systems of different natures. After reviewing spatial features in many areas it describes basics of Spatial Grasp Technology (SGT) and its Spatial Grasp Language (SGL) for dealing with large distributed systems, which can provide spatial vision, awareness, management, and even consciousness. It shows how SGL scenarios self-evolve in physical, virtual or imaginable spaces and provides details of the technology implementation, where distributed SGL interpreter integrated with advanced communication systems becomes a powerful spatial engine. Techniques are described in SGL for vision, understanding and modifying of large distributed spaces, also for security, automatic control and withstanding cyber attacks in networked systems. The book compares Spatial Grasp model with algorithms, confirming universality of the former for representing, understanding, simulation and management of any spatial and distributed systems, with the latter just representing tools for concrete applications.

1.1 Introduction

The humankind exists, operates, and develops in large distributed spaces, where "spatial" may be considered as the basic feature philosophically, conceptually, and practically. "***Spatial world***" refers to how phenomena are organized and perceived within a defined environment, encompassing concepts from geographic organization to virtual environments like the metaverse. It can describe the physical arrangement of features on Earth, mental models of our surroundings, or the structured 3D environments created by using spatial computing and virtual reality platforms. The term highlights the importance of spatial relationships, data, and experiences in understanding, interacting with, and transforming both physical and digital spaces.

The ***aim of this book*** is to analyse suitability and efficiency of the developed Spatial Grasp model and technology for dealing with spatial systems of very different

P. S. Sapaty, *Distributed Management with Spatial Grasp Model*, Studies in Systems, Decision and Control 666, https://doi.org/10.1007/978-3-032-19378-0_1

natures, show its advantages over other models and systems, and compare with the concept of algorithm.

The rest of the Chapter discusses typical spatial world problems; reviews other books in the spatial area; presents basics of Spatial Grasp Model and Technology and its Spatial Grasp Language; explains publications which influenced this book project mostly; provides summary of other book chapters; and lists publications discussed or used throughout the book, from recent to their history.

1.2 Spatial World Problems

The following can be mentioned in different areas.

Spatial development problems include inadequate integration of environmental and social aspects into planning, outdated planning systems unable to address modern issues like climate change and inequality, a lack of effective governance and public–private partnerships, insufficient funding and data.

Spatial education problems encompass problems with mental rotation and navigation which can impact learning in subjects like math and science, and broader systemic issues including inequities in resource allocation, teacher training, and a lack of standardized approaches to spatial assessment and curriculum development.

Spatial economy problems include uneven regional development, the concentration and dispersion of economic activity leading to peaks and troughs in wealth and employment, issues related to transport and transaction costs that influence location choices, and the spatial inequality experienced by those trapped in low-opportunity areas.

Spatial intelligence problems manifest as difficulties with mental visualization, navigation, and understanding spatial relationships, often seen in tasks like reading maps, solving puzzles, or judging distances. These challenges can impact everyday activities such as coordinating movement, dressing, or navigating an environment.

Spatial defence problems include adversarial actions such as GPS jamming/spoofing and anti-satellite weapons, vulnerability to cyberattacks, and the difficulty of attribution for these incidents. Other challenges involve rapid technological advancement creating an arms race, the need for resilient and proliferated satellite architectures to counter attacks.

Spatial security problems encompass vulnerabilities in space-based infrastructure and the cyber systems that support them, leading to risks of cyberattacks, physical disruptions, and the misuse of space assets for malicious purposes. Threats include destructive physical attacks like anti-satellite missile tests.

Spatial warfare problems include the physics of vast distances and high speeds, making targeting, tracking, and reactive maneuvers difficult. Significant challenges also arise from space debris, which threatens collisions and could make orbits unusable, and the environmental hazards of solar storms.

Some related references may include [1–3], but they can be numerous as almost every area of human spatial activity is not without problems, which may be essential and even dominating.

1.3 Other Books in the Area

The analyzed books [4–10], using "spatial" in their names or contents, are discussing different distributed and spatial-based ideas and solutions. Some of these appeared useful for the current book project and are mentioned below.

Fundamentals of cognition provide a basic, reader-friendly introduction to the key cognitive processes we use to interact successfully with the world around us. Our abilities in attention, perception, learning, memory, language, problem solving, thinking, and reasoning are all vitally important in enabling us to cope with everyday life. Understanding these processes through the study of cognitive psychology is essential for understanding human behaviour [4].

Spatial knowledge representation and reasoning are relevant issues for many application areas such as robotics, geographical information systems, and computer vision. Exceeding purely quantitative approaches, more recently initiated qualitative approaches allow for dealing with spatial information on a more abstract level that is closer to the way humans think and speak [5].

Complex transformations shaping societies, economies, and urban environments are discussed. This brings together interdisciplinary research on innovation, spatial development, economic policy, and digitalization, highlighting the challenges and opportunities faced by regions and industries in an era of rapid change. Also examines key issues in urban planning, regional innovation ecosystems, intercultural communication, and the evolving geopolitical landscape [6].

Redefinition of building blocks of development and existing challenges can derail child's functioning and learning. The vision and space of what is seen by the eyes, then transformed by the mind and experienced as movement, play underestimated crucial role in the development of the child's thoughts and feelings [7].

The challenges of growth in transport, complicated by scarce financial resources, landscape, and local opposition against new infrastructure investment, require new solutions. The main objective is to accelerate and jointly develop the transport capacity of the entire corridor by ensuring optimal economic benefits and spatial integration, while reducing negative impacts on the environment at local and regional level [8].

Sustainable planning and development is discussed. It covers multiple environmental issues and examines how potential solutions can and should be addressed in the context of space and place. Also offers an invaluable discussion on the need for a framework for system-based thinking where various disciplines should work together to find innovative solutions [9].

Manipulating geometric or spatial objects in the computer is common in geology, cartography, computer aided design, etc., and many of these have developed their own data structures and techniques often independently. But in many cases the developed object types and spatial queries look similarly, so attempts should be made to find a common theme [10].

1.4 The Approach Used

Towards accomplishing the book's goals we will be using the developed and tested in different countries ***Spatial Grasp Model and Technology (SGT)*** reflected by patent [11], books [12–22], latest papers [23–27], also its long history (including [28–84]). Within SGT, a high-level scenario in recursive ***Spatial Grasp Language (SGL)*** propagates and matches distributed environments in parallel wavelike mode, as symbolically depicted in Fig. 1.1.

SGT uses self-navigating, self-matching, and self-replicating recursive patterns in SGL which can directly represent complex spatial tasks and dynamically cover and impact large distributed environments. SGL top level organization can be expressed just in a single string-formula mode:

$$\boldsymbol{grasp} \rightarrow \boldsymbol{constant}|\boldsymbol{variable}|\boldsymbol{rule}(\{\boldsymbol{grasp},\})$$

An SGL scenario, or ***grasp***, initially applied in some point (or points) of distributed spaces can be a ***constant*** directly providing the result or a ***variable*** which content was or will be obtained from this or other space points. It can also be a ***rule*** expressing certain ordering, action, control, description or context over parenthesized operands, generally as ***grasps*** too. SGL can express effective, integral, and distributed solutions in large networked systems in a simple and very compact way. Space-distributed interacting SGL interpreters (potentially up to thousands and millions of copies) integrated with advanced communication systems can represent ***powerful spatial engines*** operating without central resources or control.

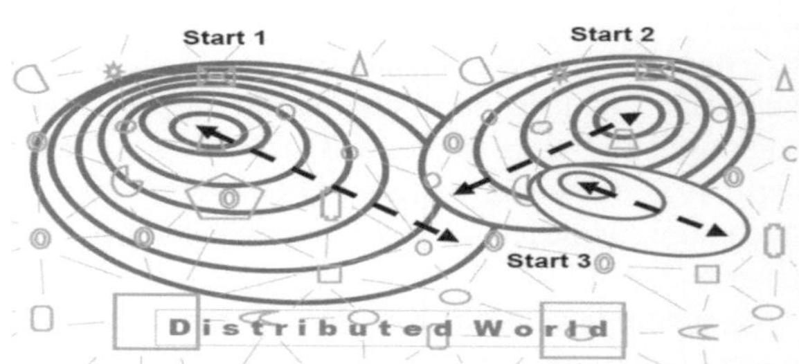

Fig. 1.1 Parallel recursive wavelike world coverage with SGT

1.5 Latest Publications Which Influenced This Book

The latest publications of the author that essentially influenced the current book can be cited as [21, 22, 24–26]. The following areas, ideas, and solutions touched in them proved particularly useful for the book project.

Spatial networking [21]. As a universal approach, networking can be used for dealing with complex problems in very different areas like physical, virtual, psychological, mental, and others. The book proposes advanced networking techniques that can simultaneously cover and manage transport networks, communication networks, social networks, battle and military networks, economic networks, distribution networks, virtual networks, neural networks, psychological networks, and criminal networks.

Global and international security [24]. Global and international security should be deeply embedded and integrated with bodies of real systems wherever in physical, virtual, or combined spaces they exist. It must be really distributed and spatial, self-organized, dynamic, and exhibiting integrity, awareness, and consciousness features. The paper offers practical security solutions for tracing forbidden items, roaming criminals, recovery from natural and human-made disasters, and restoration of damaged transport networks.

Self-healing and self-recovering systems [22, 25]. In our dynamic and extremely complex world we need measures to cope with regular emergencies, destructions, and crises. Practical solutions have been offered that can provide efficient improvement and healing in self-recovery mode, using mostly internal system resources. The developed universal self-healing procedure associated with network nodes effectively converts the whole network into "immortal spatial creature" operating any time, after any damages, and without external control.

Comparison with traditional algorithms [26]. The humankind exists, operates, and develops in large distributed spaces, where "spatial" may be considered as basic and dominant feature. The paper shows applications of SGT for dealing with spatial systems of different natures and its advantages over traditional algorithms or their extensions, usually used as specific tools for solving concrete problems, whereas Spatial Grasp model represents universal philosophy and paradigm for managing the whole world.

1.6 Summary of Other Book Chapters

Chapter 2: "Spatial" as the Basic Feature of Human Activity

"Spatial" refers to the location, arrangement, and distribution of things, including human activities, people, and resources, across the Earth's surface and within physical environments, making it a fundamental concept in geography and other fields like sociology and urban planning. Human activities, from daily journeys to the migration of populations, inherently occur within a spatial context, influencing and being influenced by physical environments and leading to distinct spatial patterns and relationships that are studied to understand human behavior, resource use, and societal development. The Chapter reviews spatial as the basic feature of various human activities with references to related sources and publications, and expresses the need for a unified philosophical, psychological, and technological approach for investigation and problem management in these and other spatial areas.

Chapter 3: Spatial Grasp Model and Technology (SGT)

Within Spatial Grasp Model and Technology (SGT) a high-level operational scenarios in recursive Spatial Grasp Language (SGL) starting in any world points effectively propagate, cover and match the distributed environment in parallel wave-like mode. Communicating SGL interpreters (up to thousands even billions copies if needed) integrated with advanced communication systems can altogether represent powerful spatial engines operating without central management and control often vulnerable in complex and unpredictable situations. The Chapter describes basics of SGT oriented from the very beginning on seeing, staying, moving, and operating in spaces of any complexity and coverage. This includes details of SGL, explanation of self-evolving recursive SGL scenarios, and implementation of the networked SGL interpreter.

Chapter 4: Spatial Grasp Language (SGL) Features

The Chapter first reviews existing distributed programming languages including Java, Python, Go, Erlang, and Scala, which are selected for comparison with Spatial Grasp Language (SGL), the key component of the chosen for this book Spatial Grasp Technology, as SGL is fundamentally oriented on distributed programming too. After providing summary of the full SGL syntax, it describes in detail the full repertoire of SGL components and their usage, which include constants, variables and universal rules. Then compares SGL with other languages by explaining their similarity and differences, and concludes the Chapter with confirmation of SGL suitability for highly integral and holistic solutions in any distributed spaces.

Chapter 5: Representing Physical and Virtual Spaces in SGL

The Chapter reviews different ideas, features, and publications related to physical, virtual and hybrid spaces, as well as their unity and integration. Representation of different features of distributed spaces is analysed in SGL, and concrete methods and mechanisms of spatial vision, penetration, navigation and management are offered in SGL. These include declaration and management of spatial communities, operations with large distributed networks and their borders, finding perimeters of complex spatial regions by collective navigation of their spaces, hierarchical network organization, as well as possible spatial art. The provided investigations and resulted solutions confirm clarity, efficiency, and compactness of expressing them in SGL.

Chapter 6: Examples of Spatial Management in Different Areas

The Chapter reviews different spatial management issues by existing publications, and then offers and explains concrete tools and scenarios for spatial management in SGL. These include finding and using shortest path tree and spatial pattern matching in networks, worldwide search for certain individuals identified by specific features and assisted by distributed security systems, chasing and fighting multiple targets at a battlefield under global awareness supported by the chasers regularly informing each other about the targets seen. Other scenarios are for tracking mobile objects by distributed radar networks, cooperative robotic finding of oil spill center, discovering, surrounding, and battling forest fires, also cooperation of autonomous vehicles within optimized lane manoeuvring.

Chapter 7: Cybersecurity Applications of SGT

Spatial security is for protection of systems based on geographical location, and cybersecurity is about protecting systems, networks, and programs from digital attacks. Implementing effective cybersecurity measures is particularly challenging today because there are more devices than people, and attackers are becoming more innovative. After reviewing different spatial security and cybersecurity issues and types of cyberattacks, the Chapter shows in SGL how to find virus sources in distributed networks by tracing infected predecessors or registered infection time in nodes, offers collective malware removing scenario using self-growing and self-organized intelligent cleaning team. It then shows how to organize massive removal of debris around the Earth, how to find and use strongest and weakest network parts, like cliques and articulation points, for improving security of distributed networks, and provides universal network recovery scenario which allows the whole network topology to self-recover after arbitrary damages.

Chapter 8: Spatial Grasp Model Versus the Concept of Algorithm

Algorithm is usually considered as a sequence of rigorous instructions for solving a class of specific problems. A review of definitions of algorithm and its modifications like parallel, distributed, spatial analysis, mobile agents too, is provided. The Chapter compares SGT principles with algorithms, where self-evolving spatial scenarios in SGL often resemble massive spatial phenomena like virus, smog, radiation, rain, smell, light, or opinion, which are far more general and global than algorithmic sequences. It then treats SGT as a self-evolving recursive pattern spatially matching networked systems, potentially any continuous spaces too, up to the whole world. And finally shows how dynamic spatial capabilities of SGT may relate to non-algorithmic ideas especially popular in relation to living organisms and societies.

Chapter 9: Conclusions

This summarizes the investigation, research, analysis, and development of this book. The obtained results can be used for the development of advanced technologies for dealing with large and complex terrestrial and celestial spaces and systems, also as radically new spatially-oriented teaching and research courses and programs at universities and academic research institutions. The described Spatial Grasp Model, Technology, and Language can be easily and quickly implemented on any platform, which was investigated and prototyped for its previous versions in different countries. Multiple SGL interpreter copies integrated with other systems and advanced communications, can altogether represent powerful spatial engines effectively solving any spatial problems and operating without central resources or control.

References

1. Cantwell, J., Shukla, P.: Spatial development of technological knowledge and the evolution of international business activity across technological paradigms. Int. Bus. Rev. **34**(1) (2025). https://ideas.repec.org/a/eee/iburev/v34y2025i1s0969593124001033.html
2. Redding, S.J.: Spatial Economics (2024). https://doi.org/10.3386/w33125
3. New Frontiers in Global Security: Emerging Technologies and Outer Space. https://unu.edu/news/new-frontiers-global-security-emerging-technologies-and-outer-space
4. Eysenck, M.W., Brysbaert, M.: Fundamentals of Cognition. Routledge (2018). ISBN-10:113867043X
5. Renz, J.: Qualitative Spatial Reasoning with Topological Information. Lecture Notes in Computer Science, 2293. Springer (2002). ISBN-10:3540433465
6. Shkutsko, A.N., Semyachkov, K.A.: Emerging Frontiers in Spatial Development, Economic Integration, and Innovation Ecosystems: Research Anthology and Selected Contributions. Independently Published (2025). ISBN-13:979-8311345224
7. Wieder, S., Wachs, H.: Visual/Spatial Portals to Thinking, Feeling and Movement: Advancing Competencies and Emotional Development in Children with Learning and Autism Spectrum Disorders. Profectum Foundation (2012). ISBN-10:0578111284
8. Drewello, H., Scholl, B.: Integrated Spatial and Transport Infrastructure Development: The Case of the European North-South Corridor Rotterdam-Genoa (Contributions to Economics). Springer (2015). ISBN-10:3319157078
9. Dixon, B.: Interdisciplinary Environmental Solutions: Using Geospatial Technologies for Bridging Disciplines, Scale and Data. Palgrave Macmillan (2025). ISBN-10 3031167627
10. Gold, C.: Spatial Context: An Introduction to Fundamental Computer Algorithms for Spatial Analysis. ISPRS Book Series 12. CRC Press (2018). ISBN-13:978-1498779104
11. Sapaty, P.S.: A distributed processing system. European Patent N 0389655, Publ. 10.11.93. European Patent Office (1993)
12. Sapaty, P.S.: Mobile Processing in Distributed and Open Environments. Wiley, New York (1999)
13. Sapaty, P.S.: Ruling Distributed Dynamic Worlds. Wiley, New York (2005)
14. Sapaty, P.S.: Managing Distributed Dynamic Systems with Spatial Grasp Technology. Springer (2017)
15. Sapaty, P.S.: Holistic Analysis and Management of Distributed Social Systems. Springer (2018)
16. Sapaty, P.S.: Complexity in International Security: A Holistic Spatial Approach. Emerald Publishing (2019)
17. Sapaty, P.S.: Symbiosis of Real and Simulated Worlds under Spatial Grasp Technology. Springer (2021)
18. Sapaty, P.S.: Spatial Grasp as a Model for Space-based Control and Management Systems. CRC Press (2022)
19. Sapaty, P.S.: The Spatial Grasp Model: Applications and Investigations of Distributed Dynamic Worlds. Emerald Publishing (2023)
20. Sapaty, P.S.: Providing Integrity, Awareness, and Consciousness in Distributed Dynamic Systems. CRC Press (2024)
21. Sapaty, P.S.: Spatial Networking in the United Physical, Virtual, and Mental World. Springer (2024)
22. Sapaty, P.S.: Self-Healing and Self-Recovering Systems Under the Spatial Grasp Model. Emerald Publishing Limited (2025)
23. Sapaty, P.S.: Spatial networks as models for organoid cultures and brain research. Mathematical Mach. Syst. (1) (2024). ISSN 1028-9763. http://www.immsp.kiev.ua/publications/articles/2024/2024_1/01_24_Sapaty.pdf
24. Sapaty, P.S.: Global and international security under spatial grasp paradigm. Int. Relat. Dipl. **12**(2), 72–85 (2024). https://doi.org/10.17265/2328-2134/2024.02.004. https://www.davidpublisher.com/index.php/Home/Article/index?id=%2050834.html

25. Sapaty, P.S.: Self-recovering-systems-under-spatial-grasp-technology. Cris. Stress. Hum. Resil. Int. J. **6**(2) (2024). https://www.crisisjournal.org/article/126490-self-recovering-systems-under-spatial-grasp-technology
26. Sapaty, P.S.: Spatial Grasp Model for Distributed Management and Its Comparison With Traditional Algorithms. Int. Relat. Dipl. **13**(3) (2025). https://doi.org/10.17265/2328-2134/2025.03.004. https://www.davidpublisher.com/Public/uploads/Contribute/689ada551780e.pdf
27. Sapaty, P.S.: Spatial management of distributed dynamic worlds. Cris. Stress. Hum. Resil. Int. J. **7**(1) (2025). crisisjournal.org
28. Bondarenko, A.T., Mikhalevich, S.B., Nikitin, A.I., Sapaty, P.S.: Software of BESM-6 computer for communication with peripheral computers via telephone channels. In: Computer Software, vol. 5. Institute of Cybernetics Press, Kiev (1970)
29. Sapaty, P.S.: A method of organization of an intercomputer dialogue in the radial computer systems. In: The Design of Software and Hardware for Automatic Control Systems. Institute of Cybernetics Press, Kiev (1973)
30. Sapaty, S.: A wave language for parallel processing of semantic networks. Comput. Artif. Intell. **5**(4), 289–314 (1986)
31. Sapaty, P.S.: Distributed modeling of cooperative behavior by mobile agents. In: Proceedings of Sixth Conference on Computer Generated Forces and Behavioral Representation. IST UCF, Orlando, FL, July 1996, pp. 599–613
32. Sapaty, P.S.: Distributed artificial brain for collectively behaving mobile robots. In: Proceedings of Symposium and Exhibition Unmanned Systems 2001, Baltimore, MD, July 31–Aug 2, 2001
33. Sapaty, P.S.: Crisis management with distributed processing technology. Int. Trans. Syst. Sci. Appl. **1**(1), 81–92. ISSN 1751-1461 (2006)
34. Sapaty, P.S.: Distributed technology for global dominance. In: Proceedings of International Conference Defense Transformation and Net-Centric Systems 2008, as part of the SPIE Defense and Security Symposium. World Center Marriott Resort and Convention Center, Orlando, FL, USA, 16–20 Mar 2008 (Also: Proceedings of SPIE, vol. 6981. In: Suresh, R. (ed.) Defense Transformation and Net-Centric Systems, 69810T, 3 Apr 2008)
35. Sapaty, P.S.: Distributed technology for global dominance. Keynote lecture. In: Proceedings of the Fifth International Conference in Control, Automation and Robotics ICINCO 2008, The Conference Proceedings, Funchal, Madeira, Portugal, 11–15 May 2008
36. Sapaty, P.S.: Human-Robotic Teaming: A Compromised Solution. AUVSI's Unmanned Systems North America, San Diego, USA, 10–12 June 2008
37. Sapaty, P.S.: High-level communication protocol for dynamically networked battlefields. In: Proceedings of International Conference Tactical Communications 2009 (Situational Awareness and Operational Effectiveness in the Last Tactical Mile). One Whitehall Place, Whitehall Suite and Reception, London, UK (2009)
38. Sapaty, P.S.: Distributed capability for battlespace dominance. In: Electronic Warfare 2009 Conference and Exhibition. Novotel London West Hotel and Conference Center, London, 14–15 May 2009
39. Sapaty, P.S.: Providing spatial integrity for distributed unmanned systems. In: Proceedings of 6th International Conference in Control, Automation and Robotics ICINCO 2009, Milan, Italy (2009)
40. Sapaty, P.S.: Distributed Technology for Global Control. Book chapter, Lecture Notes in Electrical Engineering, Informatics in Control, Automation and Robotics, vol. 37. Springer, Berlin (2009)
41. Sapaty, P.S.: Gestalt-based integrity of distributed networked systems. In: SPIE Europe Security + Defence. bcc Berliner Congress Centre, Berlin, Germany (2009)
42. Sapaty, P.S.: Remote control of open groups of remote sensors. In: Proceedings of SPIE Europe Security + Defence, Berlin, Germany (2009)
43. Sapaty, P.S.: Tactical communications in advanced systems for asymmetric operations. In: Proceedings of Tactical Communications 2010, CCT Venues, Canary Wharf, London, UK, 28–30 Apr 2010

44. Sapaty, P.S.: High-level technology to manage distributed robotized systems. In: Proceedings of Military Robotics 2010, Jolly St Ermins, London, UK, 25–27 May 2010
45. Sapaty, P.S.: Emerging asymmetric threats, Q&A session. In: Tactical Communications 2010, CCT Venues, Canary Wharf, London, UK, 28–30 Apr 2010
46. Sapaty, P.S.: High-level organization and management of directed energy systems. In: Proceedings of Directed Energy Weapons 2010, CCT, Canary Wharf, London, UK, 25–26 Mar 2010
47. Sapaty, P.S.: Unified transition to robotized armies with spatial grasp technology. In: International Summit Military Robotics, London, United Kingdom, 12–13 Nov 2012
48. Sapaty, P.S.: Distributed air and missile defense with spatial grasp technology. Intell. Control Autom. Sci. Res. **3**(2), 117–131 (2012)
49. Sapaty, P.S.: Global electronic dominance. In: 12th International Fighter Symposium, 6th–8th Nov 2012, Grand Connaught Rooms, London, UK
50. Sapaty, P.S.: Providing global awareness in distributed dynamic environments. In: International Summit ISR, London, 16–18 Apr 2013
51. Sapaty, P.S.: Ruling distributed dynamic worlds with spatial grasp technology. Tutorial at the International Science and Information Conference 2013 (SAI), London, UK, 7–9 Oct 2013
52. Sapaty, P.S.: Night vision under advanced spatial intelligence: a key to battlefield dominance. In: International Summit Night Vision 2013, London, 4–6 June 2013
53. Sapaty, P.S.: Integration of ISR with advanced command and control for critical mission applications. In: SMi's ISR Conference. Holiday Inn Regents Park, London, 7–8 Apr 2014
54. Sapaty, P.S.: Unified transition to cooperative unmanned systems under spatial grasp paradigm. Int. J. Trans. Netw. Commun. **2**(2) (2014)
55. Sapaty, P.S.: From manned to smart unmanned systems: a unified transition. In: SMi's Military Robotics. Holiday Inn Regents Park, London, 21–22 May 2014
56. Sapaty, P.S.: Unified transition to cooperative unmanned systems under spatial grasp paradigm. In: 19th International Command and Control Research and Technology Symposium, Alexandria, Virginia, 16–19 June 2014
57. Sapaty, P.S.: Distributed human terrain operations for solving national and international problems. Int. Relat. Dipl. **2**(9) (2014)
58. Sapaty, P.S.: Providing over-operability of advanced ISR systems by a high-level networking technology. In: SMI's Airborne ISR. Holiday Inn Kensington Forum, London, United Kingdom, 26–27 Oct 2015
59. Sapaty, P.S.: Distributed missile defence with spatial grasp technology. In: SMi's Military Space. Holiday Inn Regents Park, London, 4–5 Mar 2015
60. Sapaty, P.S.: Military robotics: latest trends and spatial grasp solutions. Int. J. Adv. Res. Artif. Intell. **4**(4) (2015)
61. Sapaty, P.S.: A brief introduction to the spatial grasp language (SGL). J. Comput. Sci. Syst. Biol. **09**(02) (2015). https://www.researchgate.net/publication/307744279_A_Brief_Introduction_to_the_Spatial_Grasp_Language_SGL
62. Sapaty, P.S.: Organization of advanced ISR systems by high-level networking technology. Math. Mach. Syst. (1) (2016). http://www.immsp.kiev.ua/publications/articles/2016/2016_1/01_2016_Sapaty.pdf
63. Sapaty, P.S.: Towards massively robotized systems under spatial grasp technology. J. Comput. Sci. Syst. Biol. **9**(1) (2016)
64. Sapaty, P.S.: Towards wholeness and integrity of distributed dynamic systems. J. Comput. Sci. Syst. Biol. **9**(3) (2016)
65. Sapaty, P.S.: Towards global goal orientation, robustness and integrity of distributed dynamic systems. J. Int. Relat. Dipl. **4**(6) (2016)
66. Sapaty, P.: Spatial grasp language (SGL). In: Advances in Image and Video Processing, vol. 4, no. 1 (2016). https://journals.scholarpublishing.org/index.php/AIVP/article/view/1922/1052
67. Sapaty, P.S.: Mosaic warfare: from philosophy to model to solution. Mathematical Mach. Syst. (3) (2019)

68. Sapaty, P.S.: Global network management under spatial grasp paradigm. Int. Robot. Autom. J. **6**(3), 134–148 (2020). https://medcraveonline.com/IRATJ/IRATJ-06-00212.pdf
69. Sapaty, P.S.: Global network management under spatial grasp paradigm. Glob. J. Res. Eng. J Gen. Eng. **20**(5), 58–81. Version 1.0 (2020). https://globaljournals.org/GJRE_Volume20/6-Global-Network-Management.pdf
70. Sapaty, P.: Advanced terrestrial and celestial missions under spatial grasp technology. Aeronaut. Aerosp. Open Access J. **4**(3) (2020). https://medcraveonline.com/AAOAJ/AAOAJ-04-00110.pdf
71. Sapaty, P.: Spatial management of distributed social systems. J. Comput. Sci. Res. **02**(03) (2020). https://ojs.bilpublishing.com/index.php/jcsr/article/view/2077/pdf
72. Sapaty, P.: Towards global nanosystems under high-level networking technology. Acta Sci. Comput. Sci. **2**(8) (2020). https://www.actascientific.com/ASCS/pdf/ASCS-02-0051.pdf
73. Sapaty, P.: Symbiosis of distributed simulation and control under spatial grasp technology. SSRG Int. J. Mob. Comput. Appl. (IJMCA) **7**(2) (2020). http://www.internationaljournalssrg.org/IJMCA/2020/Volume7-Issue2/IJMCA-V7I2P101.pdf
74. Sapaty, P.S.: Symbiosis of real and simulated worlds under global awareness and consciousness. In: Abstract, The Science of Consciousness (TSC) Conferences, Tucson, Arizona (2020). https://eagle.sbs.arizona.edu/sc/report_poster_detail.php?abs=3696
75. Sapaty, P.S.: Spatial grasp as a model for space-based control and management systems. Math. Mach. Syst. (1), 135–138 (2021). http://www.immsp.kiev.ua/publications/articles/2021/2021_1/Sapaty_book_1_2021.pdf
76. Sapaty, P.S.: Managing multiple satellite architectures by spatial grasp technology. Math. Mach. Syst. (1), 3–16 (2021). http://www.immsp.kiev.ua/publications/eng/2021_1/
77. Sapaty, P.S.: Spatial management of large constellations of small satellites. Math. Mach. Syst. (2) (2021). http://www.immsp.kiev.ua/publications/articles/2021/2021_2/02_21_Sapaty.pdf
78. Sapaty, P.S.: Global management of space debris removal under spatial grasp technology. Acta Sci. Comput. Sci. **3**(7) (2021). https://www.actascientific.com/ASCS/pdf/ASCS-03-0135.pdf
79. Sapaty, P.S.: Space debris removal under spatial grasp technology. Netw. Commun. Technol. **6**(1) (2021). https://www.ccsenet.org/journal/index.php/nct/article/view/0/45486
80. Sapaty, P.S.: Relation of spatial grasp paradigm to higher psychological and mental concepts. Acta Sci. Comput. Sci. **4**(12) (2022). https://actascientific.com/ASCS/pdf/ASCS-04-0359.pdf
81. Sapaty, P.S.: Seeing and managing distributed worlds with spatial grasp paradigm. Acta Sci. Comput. Sci. **4**(12) (2022). https://actascientific.com/ASCS/pdf/ASCS-04-0365.pdf
82. Sapaty, P.S.: Comprehending distributed worlds with the spatial grasp paradigm. Math. Mach. Syst. (1) (2022). http://www.immsp.kiev.ua/publications/articles/2022/2022_1/01_22_Sapaty.pdf
83. Sapaty, P.S.: Providing global awareness in distributed dynamic systems. Int. Relat. Dipl. **11**(2), 87–100 (2023). https://doi.org/10.17265/2328-2134/2023.02.002. https://www.davidpublisher.com/Public/uploads/Contribute/6486c3d05a6cc.pdf
84. Sapaty, P.S.: Simulating distributed consciousness with spatial grasp model. Math. Mach. Syst. (3) (2023). http://www.immsp.kiev.ua/publications/articles/2023/2023_3/03_23_Sapaty.pdf

Chapter 2
"Spatial" as the Basic Feature of Human Activity

Abstract "Spatial" refers to the location, arrangement, and distribution of things, including human activities, people, and resources, across the Earth's surface and within physical environments, making it a fundamental concept in geography and other fields like sociology and urban planning. Human activities, from daily journeys to the migration of populations, inherently occur within a spatial context, influencing and being influenced by physical environments and leading to distinct spatial patterns and relationships that are studied to understand human behavior, resource use, and societal development. The Chapter reviews spatial as the basic feature of various human activities with references to related sources and publications, and expresses the need for a unified philosophical, psychological, and technological approach for investigation and problem management in these and other spatial areas.

2.1 Introduction

"Spatial" is a fundamental aspect of human activity. It is influencing where and how we interact with the world. Human activity is inherently linked to space, and this spatial dimension shapes our behaviours, social structures, and even our perception of the environment. This activity covers and investigates the following:

Space as a determinant of activity where human activities are not random; they are shaped by the spatial environment;

Spatial dimensions in human actions where humans establish territories, from personal space to national boundaries;

Spatial interactions where people interact with each other and with their surroundings;

Spatial planning and design where human societies plan and design spaces to facilitate various activities;

Spatiality and social structures, where spatial organization is closely tied to social structures;

Spatial perception, where perception of space is also influenced by our activities and experiences;

P. S. Sapaty, *Distributed Management with Spatial Grasp Model*, Studies in Systems, Decision and Control 666, https://doi.org/10.1007/978-3-032-19378-0_2

Fig. 2.1 Symbolic spatial world vision

Spatial heterogeneity where human activity is not evenly distributed across space.

The spatial component is a basic feature because human actions, from daily routines to large-scale societal structures, are inherently tied to specific locations and geographical contexts. This means that understanding human behaviour requires considering where and how it occurs in space, as well as the spatial relationships between different activities and locations. The "globality" of all these issues may be symbolically expressed as by Fig. 2.1.

The aim of this chapter is to analyze and classify different areas of human existence and activity where spatial vision and understanding are most valuable and dominant, which can serve as explanation, background, and goal of this book project.

The rest of the chapter reviews different spatially-related areas supported by related publications, which include: spatial temporal analysis, space–time continuum, spatial analysis, spatial thinking, spatial vision, spatial feeling, spatial understanding, spatial awareness, spatial consciousness, spatial ability, spatial intelligence, spatial knowledge, spatial danger, spatial warfare, spatial management, spatial planning, spatial belief, spatial faith, spatial love, spatial hate, spatial crime, spatial diversity, spatial virus, spatial security, spatial patterns, and spatial art. It then expresses the need to develop a high-level universal approach capable to analyze, classify, simulate, and manage any areas and systems based on spatial nature, vision, and organization.

2.2 Review of Ideas and Activities in the Spatial World

We are starting here with ***spatial temporal analysis*** describing the ontological status of something that exists in both space and time. It is a field that uses data collected across both space and time to understand and predict phenomena. It involves studying how events or changes occur in specific locations and over specific periods, taking into account their interconnectedness. Spatio-temporal analysis involves the examination of data that has both spatial (location) and temporal (time) dimensions to describe phenomena across space and time. It's used to identify patterns, trends, and relationships in data that change over time and across different locations [1].

It also relates to ***space–time continuum*** that fuses the three dimensions of space and one of time. Because space consists of 3 dimensions, and time is 1-dimensional, space–time must, therefore, be a 4-dimensional object. It is believed to be a 'continuum' because so far as we know, there are no missing points in space or instants in time, and both can be subdivided without any apparent limit in size or duration. The space–time continuum, also known as *spacetime*, is a mathematical model that combines the three dimensions of space (length, width, height) and the one dimension of time into a single four-dimensional entity. It's a continuous fabric, meaning there are no gaps or discontinuities in space or time, and all events are located within it [2], as in Fig. 2.2.

Spatial analysis relates to knowledge, skills, and habits of mind to use concepts of space (such as distance, orientation, etc.); used as an umbrella term covering spatial perception, spatial ability, visual perception or spatial intelligence. Spatial analysis is any of the formal techniques which study entities using their topological, geometric, or geographic properties, primarily used in Urban Design. Spatial analysis includes a variety of techniques using different analytic approaches, especially spatial statistics. Key aspects of spatial analysis: *location-based data, pattern recognition, relationship analysis, decision-making* [3].

Spatial thinking is the ability to understand the relationships between objects in space, including their location, dimensions, and how they relate to each other. It's a cognitive process that involves visualizing, interpreting, and reasoning about spatial information to understand our world and develop solutions. The term *spatial thinking* is used as a unifying, common umbrella term for the human ability to direct optical stimuli received by the eye into the brain, to be able to interpret these stimuli, to be able to recognize spatial objects, to be able to mentally imagine spatial objects, and so on [4].

Spatial vision refers to the ability of an eye to capture and process fine details of a visual scene, influenced by factors such as the quality of the optical system and the density of photoreceptors on the retina. It determines the resolution and clarity of images perceived by an organism. Spatial vision, also known as *spatial resolution or spatial acuity*, refers to the ability of an eye to perceive details and separate objects close together. It's the quality of a visual system's ability to process and represent

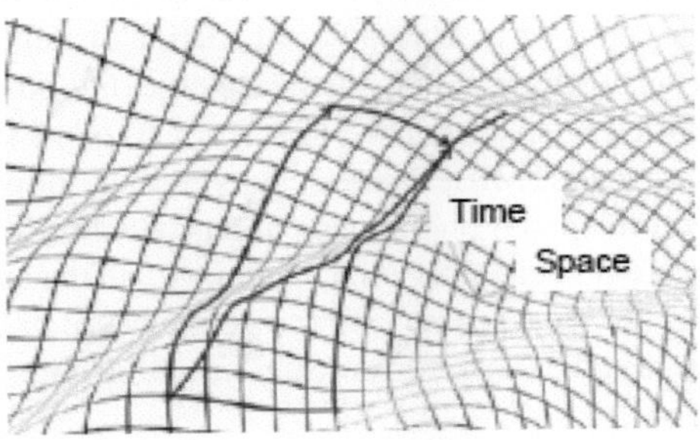

Fig. 2.2 Space–time continuum

Fig. 2.3 Space feelings

spatial patterns of light, contributing to object recognition and scene interpretation [5].

Spatial feeling uses rich illustrations and examinations of art, technology, and philosophy to explain this phenomenon. It generally refers to an awareness and understanding of one's position and surroundings within space. It's a subjective experience that involves both cognitive and sensory processes, creating a sense of "being there" or a feeling of being immersed in a particular environment. This feeling is crucial for navigation, object manipulation, and understanding the world around us [6], Fig. 2.3.

Spatial understanding is the ability to comprehend and interpret the location, size, and relationships of objects and spaces, both in the physical world and in our minds. It involves visualizing, reasoning about, and manipulating objects and scenes in a 3D environment. Spatial understanding often involves visual processing and the ability to interpret visual information in the environment, such as distance, depth, and patterns. It includes the ability to mentally rotate or manipulate objects in one's mind's eye, a skill used in tasks like stacking plates or fitting a box through a doorway. Understanding spatial relationships means knowing how objects are positioned relative to each other and to the viewer, especially across different orientations [7].

Spatial awareness is the ability to be aware of your surroundings and where you are in relation to the surrounding objects, as well as the ability to understand your body's position in relation to your surroundings. Spatial awareness is the ability to understand the relationships between objects and your own body in space, including how things move and interact within that space. It involves perceiving objects in relation to each other and oneself, and includes skills like perspective-taking, visualizing, and path-finding. Spatial awareness is thinking about objects in different spatially oriented ways. Examples of spatial awareness include: being able to do mental rotations in your head, visualizing objects from different perspectives, coordinating how different space is used in relation to other space [8].

Spatial consciousness may refer to individual or collective awareness about real-world spatial phenomena and processes; it relates to the mental process that allows us to perceive, organize, and navigate the space around us. Spatial consciousness, in its broadest sense, refers to being aware of one's position and surroundings in space and understanding how these relationships influence one's actions and interactions.

Fig. 2.4 Spatial consciousness

It goes beyond basic spatial awareness by incorporating a deeper understanding of how space impacts individuals, society, and the environment [9], Fig. 2.4.

Spatial ability is the capacity to understand, reason, and remember the visual and spatial relations among objects or space. It involves perceiving, visualizing, and mentally manipulating objects in space, and it plays a crucial role in various aspects of life, including navigation, STEM fields, and even everyday tasks. Spatial ability consists of spatial perception (orientation to one's own body), mental rotation (the ability to mentally rotate 2-dimensional or 3-dimensional objects), and spatial visualization (complex multi-step spatial processing to create a mental representation of an object) [10].

Spatial intelligence is the concept of being able to successfully perceive and derive insight from visual data. Also known as spatial reasoning, it is the ability to understand and manipulate visual information, including shapes, colours, and relationships in three dimensions. It's the capacity to perceive and make sense of the world around us visually and mentally, and it's often linked to areas like art, architecture, and engineering. Spatial intelligence provides the ability to imagine transformations of the orientation of objects (e.g., mental rotations) and the ability to imagine the consequence of observer movements around arrays of objects (i.e., perspective taking) [11], Fig. 2.5.

Spatial knowledge refers to chunks of knowledge that are focused on the spatial component like coordinates and place names. It includes factual statements about locations and shapes, as well as spatial relations between separate units. Spatial knowledge encompasses the understanding of where objects are in space and how

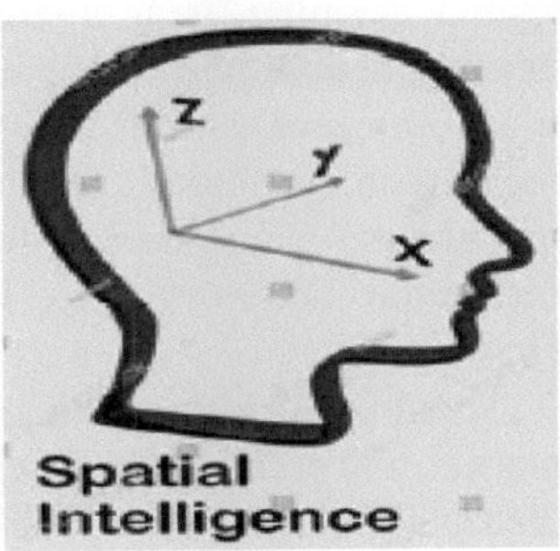

Fig. 2.5 Spatial intelligence

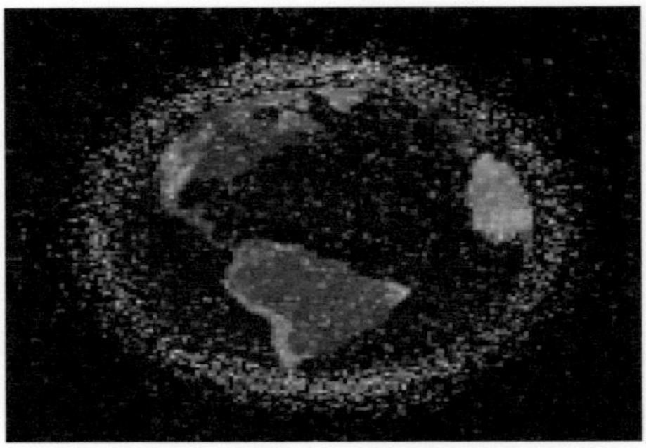

Fig. 2.6 Spatial danger

they relate to each other, including their location and orientation. It's a crucial aspect of cognition, enabling navigation, object manipulation, and spatial reasoning. This knowledge can be represented in various ways, including mental maps, cognitive graphs, and coordinate systems [12].

Spatial danger originates from the risk that an infection in this space can spread easily to either side. "Spatial danger" refers to a danger that is associated with a specific location or area, rather than being limited to a single point. It can be the result of various natural or human-induced events, such as natural disasters, environmental hazards, or military activities. It refers to a danger that is geographically or spatially related. This can include various contexts like risk assessment, superspreading environments, military context, outer space context, anatomical context, cultural heritage, and others [13], Fig. 2.6.

Spatial warfare (or "space warfare") includes ground-to-space warfare, space-to-space warfare, and space-to-ground warfare. Space warfare is combat in which one or more belligerents are in outer space. The scope of space warfare includes *ground-to-space warfare*, such as attacking satellites from the Earth; *space-to-space warfare*, such as satellites attacking satellites; and *space-to-ground warfare*, such as satellites attacking Earth-based targets. There exist international treaties, which are in place to attempt to regulate conflicts in space and limit the installation of space weapon systems, especially nuclear weapons [14].

Spatial management highlights the need for more interdisciplinary, strategic, and collaborative methods to achieve broad goals. Spatial management, also known as spatial planning, is an interdisciplinary field that deals with the arrangement and development of physical space to achieve sustainable and harmonious development. It encompasses various aspects, including land use, infrastructure, and the environment, with the goal of creating high-quality living conditions and protecting natural resources. It's an interdisciplinary field that considers social, economic, and environmental factors to shape land use, urban development, and resource allocation [15].

Spatial planning mediates between the respective claims on space of the state, market, and community. It is a public sector function aimed at influencing the spatial distribution of activities to achieve a rational organization of land use, balance development demands with environmental protection, and meet social and economic objectives. Spatial planning involves organizing human activities in geographic areas, considering various sector interests and concerns to facilitate integration and identify potential conflicts. It is an interdisciplinary field in the engineering and social sciences that deals with the spatial arrangement and development [16].

Spatial belief is a central and researched social category, educational too; it extends the cognitive research on human belief to the area of spatial reasoning. Spatial beliefs refer to how individuals understand and reason about the spatial relationships between objects or places in the world. This includes how they mentally represent and manipulate these spatial arrangements, and how they revise those beliefs when presented with new or conflicting information. It can also refer to a specific type of belief, like the belief that spatial intelligence is innate and cannot be developed, or to the use of beliefs in spatial reasoning and decision-making [17].

Spatial faith has a long history from its roots in the ancient drafting of religious cosmologies; also, astronauts observed their religions while in space, sometimes publicly, sometimes privately. It refers to the relationship between religious belief and the physical or social space where that belief is practiced or experienced, examines how space shapes religious practices, communities, and the meaning of faith itself, and how faith in turn shapes the understanding and use of space. Its key aspects include sacred and profane spaces, impact of location, contestation and conflict, religious identity and belonging, and others [18], Fig. 2.7.

Spatial Love explores the relationship between romantic love and the physical spaces it occupies and shapes. It examines how love's impact is felt in both personal and public spaces, and how these spaces in turn influence romantic relationships. Love is often thought of as an emotion, a feeling that resides within the heart and mind. But love is not just an abstract sentiment. It is something that is shaped by geography in ways both seen and unseen. It is engraved into the places we live and carried within us long after we change those places. Love, in the same way, shapes the spaces around it. The geography of love can be also defined by separation [19].

Spatial hate is examined in different perspectives and variety of spatial and temporal contexts. Examining hate crime from a spatial and temporal perspective allows us to understand it better. "Spatial hate" refers to the geographical concentration and patterns of hate speech and hate crimes. Studies have shown that hate crimes tend to be spatially concentrated, with certain areas experiencing a higher incidence of these crimes. Analyzing the spatial distribution of hate speech can reveal clusters

Fig. 2.7 Spatial faith

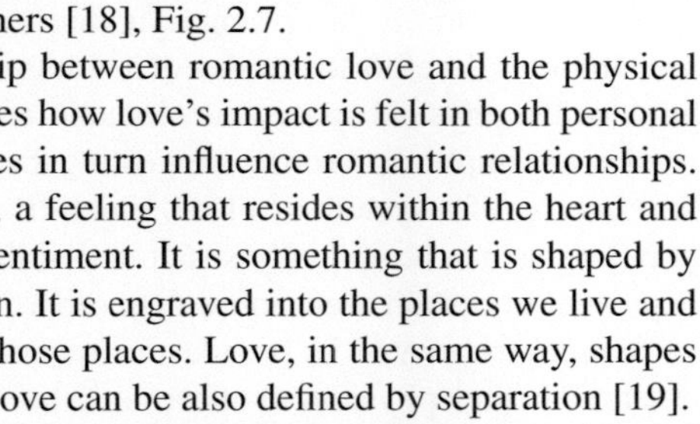

and patterns of hate language in specific regions. It also encompasses the impact of hate on how individuals and communities experience public spaces [20].

Spatial crime explains that many crime incidents may happen at particular spatial locations and points in time. Spatial criminology is a field that analyzes the spatial aspects of crime, including where and how crime occurs, the patterns of crime, and the relationship between crime and the physical environment. It focuses on understanding the distribution of crime across space and the impact of environmental factors on crime rates. Also examines how crime is distributed across a landscape, including how certain places become "crime hot spots," how crime changes over time, and how crime can be influenced by the physical and social environment [21].

Spatial diversity in a general sense refers to the variation or differences observed across geographic locations or space. Specifically, in wireless communication, it's a technique that uses multiple antennas to improve signal quality and reliability by utilizing different transmission paths and mitigating fading. Spatial diversity in Free Space Optics networks utilizes multiple apertures (transmitters and/or receivers) to mitigate the effects of atmospheric turbulence and improve signal quality. By using multiple independent channels, the technique can overcome limitations on transmit power, reduce blockage by obstacles, and increase link reliability [22].

Spatial virus is based on exploitation of multiple scales in space and time, whereas many investigations describing dynamics of infectious are based on spatially structured models. It refers to the idea of a virus spreading and affecting not just individuals but also the physical spaces they inhabit, and even the broader environment. Can be used in a metaphorical sense, suggesting that a virus can disrupt or change the normal functioning of a system or place, much like a biological virus affects a body. Viral populations can be spatially structured at the between-host level, meaning the virus is present in certain areas more than others [23], Fig. 2.8.

Spatial security in its broadest sense encompasses the measures taken to protect assets, whether physical or informational, based on their geographic location and related vulnerabilities. It involves understanding how geographical context impacts security and implementing appropriate measures to mitigate risks. This can range from securing physical infrastructure and assets based on their location, to protecting spatial data and systems from cyber threats. Spatial security covers cybersecurity and also includes the security of outer space and related technologies [24].

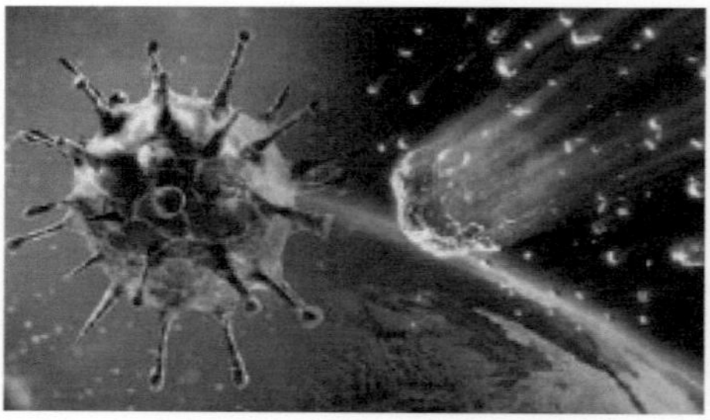

Fig. 2.8 Spatial viruses

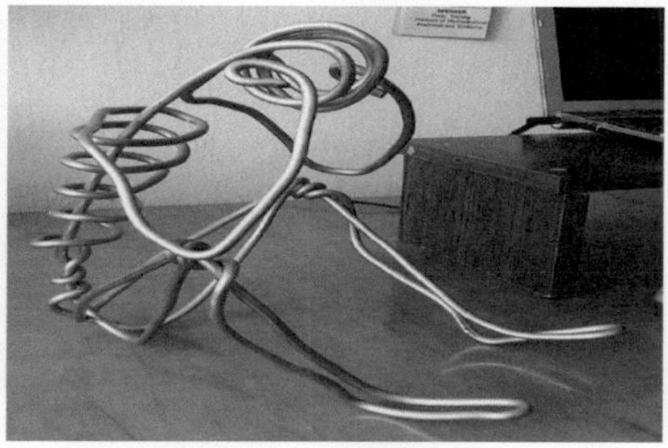

Fig. 2.9 Spatial art

Spatial patterns refer to the arrangement and distribution of features across a geographic area. They describe how things are organized on the Earth's surface, including their location, density, and relationships to each other. Understanding spatial patterns is crucial in various fields like geography, ecology, and urban planning, as it helps reveal underlying processes and trends. Examples of spatial patterns: *clustered/aggregated* with features grouped closely together; *dispersed/even/uniform*, like distribution of shopping malls; *random*, with features distributed randomly; *linear*, arranged along a line or path; *grid*, arranged in a rectangular or square pattern; and others possibly combining the mentioned above characteristics [25].

Spatial art encompasses art forms that engage with three-dimensional space, often creating environments or installations. It includes a variety of disciplines like sculpture, ceramics, performance art, and installation art. Spatial design, closely related, focuses on the design of human environments, particularly interiors, to enrich the experience of those within them. It is a broad term for art that utilizes and manipulates three-dimensional space, often creating installations or environments. Space, form, material, colour, texture, light, scale, and context are all important considerations in spatial art [26], Fig. 2.9.

2.3 Global Spatial Approach Needed

A universal concept, approach, and method are needed to effectively represent, analyze, classify, simulate, and manage areas and systems of the spatial nature. The believably appropriate paradigm, based on the developed Spatial Grasp Model, Technology, and Language [27–43], already tested in various areas, including very large, highly distributed, and dynamic, will be discussed in detail and practically used throughout the rest of this book, Fig. 2.10.

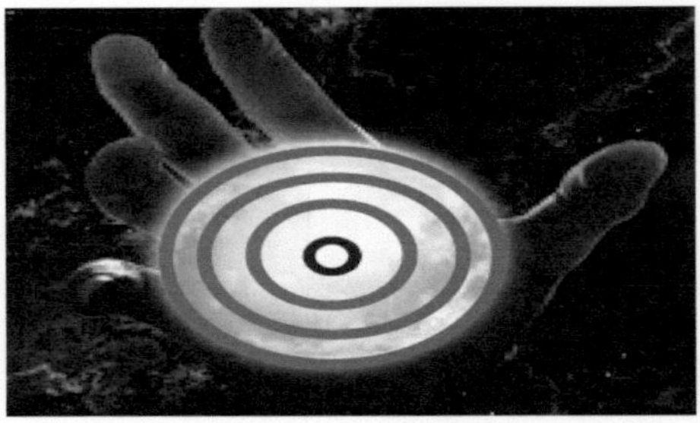

Fig. 2.10 Universal spatial grasp approach

2.4 Conclusions

The Chapter reviewed different areas with references to the related information sources and publications, where "spatial" is the dominant feature of human activities. It also expressed the need for a unified theoretical and practical approach directed to investigation and system management in these and other spatial-based areas.

References

1. Oliver, S.: What is spatial temporal in philosophy? (2019). https://www.quora.com/What-is-spatial-temporal-in-philosophy#:~:text=This%20is%20an%20adjective%20describing,as%20your%20thoughts%20and%20emotions
2. Spacetime. https://en.wikipedia.org/wiki/Spacetime
3. Spatial analysis. https://en.wikipedia.org/wiki/Spatial_analysis
4. Spatial thinking. https://geometriedidaktik.at/en/spatial-ability/
5. Spatial vision. https://www.sciencedirect.com/topics/neuroscience/spatial-vision
6. Bardt, C.: The Feeling of Space. The MIT Press (2024). ISBN: 9780262049368. https://mitpress.mit.edu/9780262049368/the-feeling-of-space/
7. Spatial understanding. https://www.sciencedirect.com/topics/computer-science/spatial-understanding
8. Seladi-Schulman, J.: What's important about spatial awareness? 21 Apr 2020. https://www.healthline.com/health/spatial-awareness
9. Galland, D., Grønning, M.: Spatial consciousness, 15 Apr 2019. https://doi.org/10.1002/9781118568446.eurs0308
10. Spatial ability. https://en.wikipedia.org/wiki/Spatial_ability
11. Fitzgibbons, L.: Spatial intelligence. https://www.techtarget.com/whatis/definition/spatial-intelligence
12. Laurini, R., Thompson, D.: 17—Spatial knowledge: intelligent spatial information systems. Fundam. Spat. Inf. Syst. 620–670 (1992). https://www.sciencedirect.com/science/article/abs/pii/B9780080924205500220
13. Danger space. https://en.wikipedia.org/wiki/Danger_space
14. Space warfare. https://en.wikipedia.org/wiki/Space_warfare
15. Dandoulaki, M., Lazoglou, M.: Disaster risk management and spatial planning: evidence from the fire-stricken are of Mati, Greece. Sustainability **15**(12), 9776 (2023). https://doi.org/10.3390/su15129776
16. Spatial planning. https://en.wikipedia.org/wiki/Spatial_planning

17. Nejasmic, J., Bucher, L., Knauff, M.: Grounded spatial belief revision. Acta Physiol. (Oxf.) **157**, 144–154 (2015). https://doi.org/10.1016/j.actpsy.2015.02.008
18. Corrigan, J.: Spatiality and religion. In: The Spatial Turn. Routledge (2008). https://doi.org/10.4324/9780203891308
19. Bielińska, A.: Love in the outer space. On the spatiality of being together. https://www.taylorfrancis.com/chapters/edit/10.4324/9781003436188-7/love-outer-space-agata-bieli%C5%84ska
20. Hopkins, P.: Afterword: Spatializing Hate—Relational, Intersectional and Emotional Approaches. Bristol University Press. https://doi.org/10.56687/9781529215205-017
21. Hipp, J.R.: The Spatial Scale of Crime, 1st edn. Routledge (2022). https://www.amazon.com/Spatial-Scale-Crime-John-Hipp/dp/103220236X
22. Spatial diversity. In: Wireless Communications Design Handbook (1998). https://www.sciencedirect.com/topics/engineering/spatial-diversity
23. Funk, G.A., Jansen, V.A.A., et al.: Spatial models of virus-immune dynamics. J. Theor. Biol. **233**(2) (2005). https://www.sciencedirect.com/science/article/abs/pii/S0022519304004758
24. Space network security, chapter. First Online: 01 Mar 2023. https://doi.org/10.1007/978-981-99-1125-7_5
25. Scott, L.M.: Analysis of spatial pattern. In: International Encyclopedia of the Social & Behavioral Sciences, 2nd edn., pp. 178–184 (2015). https://doi.org/10.1016/B978-0-08-097086-8.72064-2
26. Spatial art history tutorials and training. Arts & Sciences at Washington University in St. Louis. https://arthistory.wustl.edu/spatial-art-history-tutorials
27. Sapaty, P.S.: A distributed processing system. European Patent N 0389655, Publ. 10.11.93. European Patent Office
28. Sapaty, P.S.: Mobile Processing in Distributed and Open Environments. Wiley, New York (1999)
29. Sapaty, P.S.: Ruling Distributed Dynamic Worlds. Wiley, New York (2005)
30. Sapaty, P.S.: Managing Distributed Dynamic Systems with Spatial Grasp Technology. Springer (2017)
31. Sapaty, P.S.: Holistic Analysis and Management of Distributed Social Systems. Springer (2018)
32. Sapaty, P.S.: Complexity in International Security: A Holistic Spatial Approach. Emerald Publishing (2019)
33. Sapaty, P.S.: Symbiosis of Real and Simulated Worlds under Spatial Grasp Technology. Springer (2021)
34. Sapaty, P.S.: Spatial Grasp as a Model for Space-based Control and Management Systems. CRC Press (2022)
35. Sapaty, P.S.: The Spatial Grasp Model: Applications and Investigations of Distributed Dynamic Worlds. Emerald Publishing (2023)
36. Sapaty, P.S.: Providing Integrity, Awareness, and Consciousness in Distributed Dynamic Systems. CRC Press (2024)
37. Sapaty, P.S.: Spatial Networking in the United Physical, Virtual, and Mental World. Springer (2024)
38. Sapaty, P.S.: Self-Healing and Self-Recovering Systems under the Spatial Grasp Model. Emerald Publishing Limited (2025)
39. Sapaty, P.S.: Spatial networks as models for organoid cultures and brain research. Mathematical Mach. Syst. (1) (2024). ISSN 1028-9763. http://www.immsp.kiev.ua/publications/articles/2024/2024_1/01_24_Sapaty.pdf
40. Sapaty, P.S.: Global and international security under spatial grasp paradigm. Int. Relat. Dipl. **12**(2), 72–85 (2024). https://doi.org/10.17265/2328-2134/2024.02.004. https://www.davidpublisher.com/index.php/Home/Article/index?id=%2050834.html
41. Sapaty, P.S.: Self-recovering-systems-under-spatial-grasp-technology. Cris. Stress. Hum. Resil. Int. J. **6**(2) (2024). https://www.crisisjournal.org/article/126490-self-recovering-systems-under-spatial-grasp-technology

42. Sapaty, P.S.: Spatial grasp model for distributed management and its comparison with traditional algorithms. Int. Relat. Dipl. **13**(3) (2025). https://doi.org/10.17265/2328-2134/2025.03.004. https://www.davidpublisher.com/Public/uploads/Contribute/689ada551780e.pdf
43. Sapaty, P.S.: Spatial management of distributed dynamic worlds. Cris. Stress. Hum. Resil. Int. J. **7**(1) (2025). crisisjournal.org

Chapter 3
Spatial Grasp Model and Technology (SGT)

Abstract Within Spatial Grasp Model and Technology (SGT) a high-level operational scenarios in recursive Spatial Grasp Language (SGL) starting in any world points effectively propagate, cover and match the distributed environment in parallel wavelike mode. Communicating SGL interpreters (up to thousands even billions copies if needed) integrated with advanced communication systems can altogether represent powerful spatial engines operating without central management and control often vulnerable in complex and unpredictable situations. The Chapter describes basics of SGT oriented from the very beginning on seeing, staying, moving, and operating in spaces of any complexity and coverage. This includes details of SGL, explanation of self-evolving recursive SGL scenarios, and implementation of the networked SGL interpreter.

3.1 Introduction

The Spatial Grasp Model and Technology (SGT) is a paradigm for managing and controlling large, distributed systems by using active, self-evolving patterns written in a high-level Spatial Grasp Language (SGL). Instead of traditional programs, SGT employs patterns that self-propagate, self-replicate, and self-modify, enabling operations across physical, virtual, and executive worlds without a central server.

The aim of this Chapter is to describe main features of Spatial Grasp Technology (SGT) and its basic Spatial Grasp Language (SGL), including their implementation and operation in distributed spaces.

The rest of the Chapter describes SGT main features, including types of worlds it can operate with; provides SGL basics allowing us to create recursive self-evolving scenarios; explains how SGL scenarios spread and operate in distributed spaces in parallel forward–backward mode; describes details of spatial technology implementation including SGL interpreter structure and operation of the interpretation network; explains mechanisms of spatial control of distributed SGT-based systems using self-optimized tracking structures; concludes presentation with repeating key SGT features and also cites the used references.

P. S. Sapaty, *Distributed Management with Spatial Grasp Model*, Studies in Systems, Decision and Control 666, https://doi.org/10.1007/978-3-032-19378-0_3

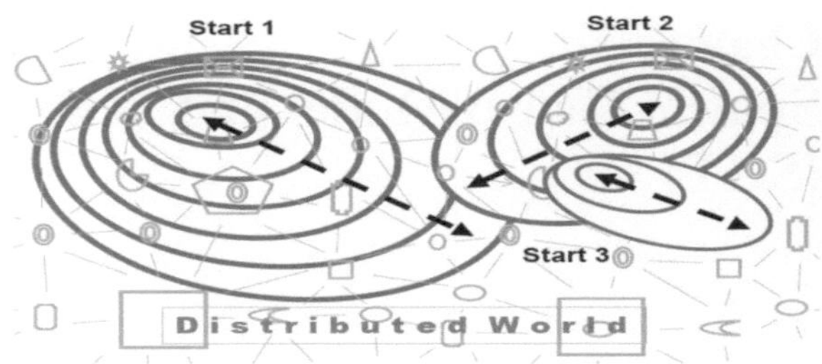

Fig. 3.1 Symbolic views of wavelike world coverage under SGT

3.2 SGT Main Features

Within Spatial Grasp Technology (SGT) (see the related patent [1], published books [2–12], latest papers [13–15], some history [16–81]) a high-level operational scenario in recursive ***Spatial Grasp Language*** (SGL), starting in any world points, propagates, covers and matches the distributed environment in parallel wavelike mode, as symbolically shown in Fig. 3.1. Such propagation can result in returning and analyzing the reached states and data as well as in further waves from the initial and new nodes which may be in any numbers.

This concept is based on quite different philosophy and practice of traditional dealing with large distributed and parallel systems. Instead of representing systems and solutions in them in the form of communicating parts or agents, the Spatial Grasp paradigm is organizing everything by integral, holistic, and parallel substance self-propagating thru and self-covering-matching distributed worlds of different natures. The latter may include:

Physical World (PW), as continuous, with points identified by physical coordinates;
Virtual World (VW), as discrete, of named nodes and named links between them;
Executive World (EW), as of active doers (like humans, robots, etc.) and communications between them.

Different combinations of these worlds can also be possible within the same paradigm, like Virtual-Physical World (VPW), Virtual-Executive World (VEW), Executive-Physical World (EPW), and Virtual-Executive-Physical World (VEPW) combining features of the previous cases.

3.3 Spatial Grasp Language Basics

The recursive language top level organization can be expressed just in a single string-formula mode:

***grasp* → *constant* | *variable* | *rule* ({ *grasp*, })**

which can be further extended as follows:

constant → information | matter | special | custom
variable → global | heritable | frontal | nodal | environmental
rule → movement | creation | advancement | branching | cycling |
echoing | verification | assignment | transference |
exchange | timing | qualifying | type | usage

Some details on its components as follows.

Constant can be self-identifiable by the way written or defined by special rules embracing arbitrary textual representations, with constants being as follows:

Information—As arbitrary strings in single quotes (which may be nested, if multiple), also numbers in traditional representations;
Physical matter or objects—As any texts in double quotes (which can be nested too) adequately identifying these elements;
Special or reserved items—As standard parameters or modifiers for different language rules;
Custom items—As self-identifiable names/words established additionally to serve specific classes of problems.

Variable types can be defined by how their names are written, but in general, variables may have any names if their types are declared by special rules; with details of their types as follows.

Global variables (most expensive and rarely used) which can simultaneously serve any SGL scenarios;
Heritable variables appearing in the reached world points and serving all subsequent scenario steps;
Frontal variables as mobile and transferred between subsequent scenario steps together with moving control;
Nodal variables associated with the reached space positions and shared with any scenarios in the same positions;
Environmental variables accessing parameters of navigated worlds, some internal interpretation features too.

Rule may be detailed as follows.

Movement may result in virtual hopping to the existing nodes (having virtual or/and executive dimensions) or in real movement to new physical locations, subsequently starting the remaining scenario in the nodes reached.
Creation creates or removes nodes and/or links leading to them during distributed world navigation; after termination, the resultant values correspond to the names and positions of the reached nodes, and the next scenario steps may start from all these nodes.
Advancement can organize forward advancement in space and time of the embraced scenarios, i.e. "in depth", which can evolve within their sequence in

synchronous or asynchronous manner, with resultant positions reached by the final stage.

Branching rules allow the embraced set of scenario operands to develop "in breadth", each from the same starting position, with the resultant set of positions and order of their appearance depending on the logic of a concrete branching rule.

Cycling repeatedly invokes the embraced scenario; the resultant set of positions on the rule will be integration of all positions after successful scenario invocations from the same point, or those obtained on its last invocation in a sequence, depending on the type of cycling.

Echoing oriented on various aspects of data and knowledge processing containing rules which may use local or remote values for different operations which should be processed in the starting points.

Verification rules verify the result of concrete procedure while remaining after their completion in the same world positions where they started.

Assignment rules assign the result of the right scenario operand, which may be arbitrarily remote, to the variable or set of variables named or reached by the left scenario operand, which may be remote too.

Transference rules organize transference of control in distributed scenarios to the code treated either as SGL procedure or an external system; they can also trigger broadcasting of the data obtained to different destinations.

Exchange rules provide input of external information or physical matter by the initiative of SGL scenario, also output of the resultant value obtained by the embraced scenario outside or to other nodes.

Timing rules are dealing with conditions related to a time allowed for the scenarios they embrace, or provide the needed time delays between different scenario parts.

Qualifying rules are providing or supporting certain qualities or abilities, also setting constraints or restrictions to the scenarios they embrace.

Type rules explicitly assign types to different constructs, using their existing repertoire.

Usage rules explain how to use the information units they embrace.

3.4 Self-Evolving SGL Scenarios

The following are some hints on how SGL scenarios self-evolve in distributed environments.

They are developing in ***steps, potentially parallel***, with new steps produced on the results of previous steps. Any step is associated with a certain point or points of the world. Each step provides a resultant value (single or multiple) and resultant control state.

Different scenario parts may evolve from the same points in ***ordered, unordered, or parallel manner***, as independent or interdependent branches. Different scenario parts can spatially succeed each other, with new parts evolving from positions reached by the previous parts.

This potentially parallel and distributed scenario can proceed in ***synchronous or asynchronous modes***, also any combinations. SGL operations and decisions can use control states and values returned from subsequent scenario parts, effectively combining controlled forward and backward scenario evolution.

Different steps from the same or different scenarios may happen to be temporarily associated with the ***same world points***, while sharing persistent or provisional information in them. Staying with different world points, the scenarios can analyze and impact the navigated worlds via these locations.

Scenarios navigating distributed spaces can form active ***distributed infrastructures*** in them, which can be shared with other scenarios. Overall organization of the world creation, coverage, modification, analysis and processing provided by SGL rules can be arbitrarily nested.

The evolving SGL scenarios can abandon utilized parts if not needed any more, also self-modify and self-replicate during space navigation. The special control states (***thru***, ***done***, ***fail***, ***fatal***) appear after completion of different scenario steps, indicating their progress or failure. They can be used for effective control of multiple distributed processes with proper decisions at different levels.

3.5 Spatial SGT Implementation

- **SGL interpretation network**

Each SGL interpreter copy can handle and process multiple active scenario code propagating in space and time. Communicating interpreters can be of arbitrary number of copies (up to thousands and millions if needed) effectively integrated with other existing systems and communications, altogether representing ***powerful spatial engines*** operating without central resources or control. Hardware or software SGL interpreters, shown in Fig. 3.2 as universal control and processing units (working with complex spatial graph and network data) can be installed, runtime created too, in proper physical or virtual world positions.

- **SGL interpreter organization**

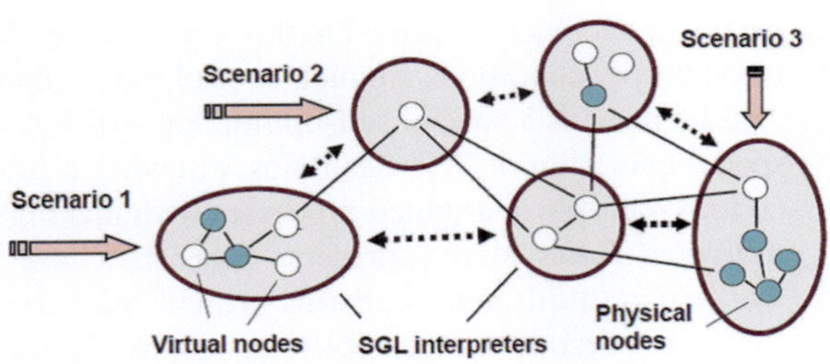

Fig. 3.2 Distributed SGL interpretation working with complex spatial data

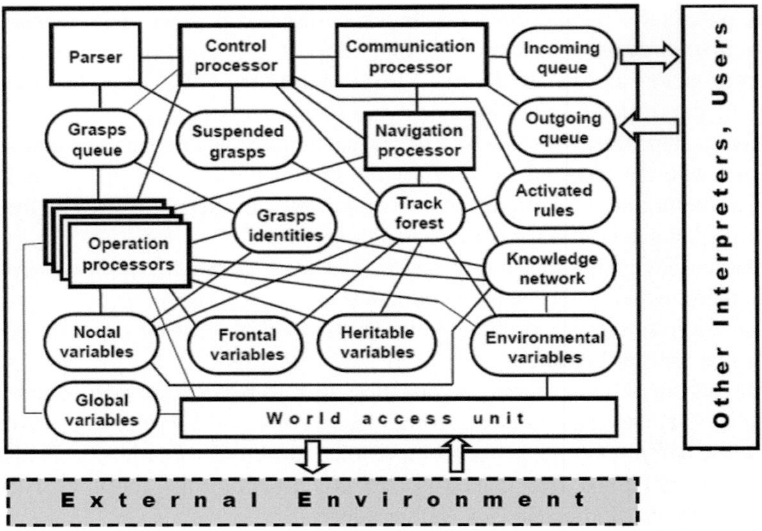

Fig. 3.3 SGL interpreter components and their interactions

The SGL interpreter main components and its general organization are shown in Fig. 3.3.

The interpreter consists of a number of specialized functional processors (shown by rectangles) working with and sharing specific data structures. These include: ***Communication Processor***, ***Control Processor***, ***Navigation Processor***, ***Parser***, different ***Operation Processors***, and special (external and internal) ***World Access Unit*** directly manageable from SGL. The interpreter's basic data structures hold permanent, persistent, and/or temporary information about the navigated and processed worlds, as follows.

Grasps Queue keeps multiple scenario fragments (syntactically represented as *grasps*) which are parsed and ready for execution in this particular interpreter. Independent from each other and queued by the priority or arrival time, they can be processed sequentially or in parallel.

Suspended Grasps structure keeps deferred parts of SGL scenarios which will be parsed in detail and executed in this interpreter or forwarded to other interpreters when proper conditions are met, control states and/or data availability including.

Track Forest is a special self-optimizing structure reflecting *dynamic history* of spatial evolution of SGL scenarios, allowing it to automatically coordinate and control multiple distributed processes with making decisions at different levels.

Activated Rules store represents rules that have been activated and continue working with different scenarios or their branches. Some rules may be waiting for their operands to be completed (say, for obtaining data to be processed or the echoed control states to make decisions).

Knowledge Network store keeps interlinked data on parts of virtual, physical, execution, or combined worlds associated with the current interpreter, this including named or unnamed nodes (which may have physical coordinates) and named links connecting them.

Grasps Identities are keeping personal colours, or identities, of scenarios that have been currently processed in this interpreter (in an active or suspended way), directing them to the resources related to these scenarios, like nodal variables and knowledge network elements.

Heritable Variables store keeps all heritable variables with their contents created by SGL scenarios within the current interpreter, these variables being linked to the track forest nodes within which they originated.

Frontal Variables store holds all frontal variables with their contents registered at the current moment of time within the current interpreter. These variables are linked with the fringe nodes of evolving track trees reflecting space–time evolution of generally hierarchical and parallel processes.

Nodal Variables store keeps all nodal variables with their contents which are linked to nodes in the Knowledge Network store, being their temporary property until the scenarios that created them remain alive in the distributed space (not necessarily in the current interpreter only).

Environmental Variables store keeps special variables with reserved names, which allow us to control and work directly with different space and time features of the worlds created, navigated and processed, also with internal parameters of the interpreter.

Global Variables store keeps information on global variables, being the most expensive ones, which can be simultaneously used and shared by any scenarios or their different branches, and at any time.

Incoming Queue accepts and stores incoming messages, which may be complete SGL scenarios from users or their parts coming from other interpreters, to be executed in the current interpreter and, possibly, continued in other interpreters afterwards.

Outgoing Queue store accumulates obtained results to be passed to the directly connected users, also parts of SGL scenarios with accompanying intermediate data to be forwarded to other interpreters for further consideration.

3.6 Spatial Control of SGT

As both backbone and nerve system of the distributed interpreter, its self-optimizing ***Spatial Track System*** provides hierarchical command and control as well as remote data and code access. It also supports spatial variables and merges distributed control states for decisions at higher organizational levels. The track infrastructure is automatically distributed between active components (humans, robots, computers, smartphones, satellites, etc.) during scenario self-spreading in distributed environments. It integrates the following operational stages.

Forward grasping. In the forward process, next steps of scenario development are forming new track nodes connected to the previous nodes by track links. Reflecting the history of scenario evolution, this growing track structure is effectively supporting heritable, nodal and frontal variables associated with track nodes.

Echoing. After completing the forward stage of SGL scenario, the distributed track system can provide and return to the starting track node the generalized control state based on termination states in all fringe nodes. The track system, on the request of higher-level scenario rules, can also collect the local data obtained at its fringe nodes and process them hierarchically and in parallel, or just merge them into a resultant list of values echoed to the higher levels. The track echoing process also optimizes the track system by removing already used but not needed any more nodes, reducing distance from the start to fringe positions.

Further forward development. The echo-modified and optimized track system can route further grasps to the world positions reached by the previous grasps, which may be to all or only part of them depending on higher-level decisions and operations. Heritable variables created in certain track nodes can be accessed from these and all subsequent positions in the track tree system (for both reading and writing operations).

Some of these stages are symbolically exhibited in Fig. 3.4 and explained below.

(a) Starting with a single track node 1, with fixed heritable variable H and mobile frontal variable F in it;
(b) Track growing and splitting with moving and replication of frontal variable F in nodes 2 and 3, also creation of fixed nodal variable N in node 3;
(c) Further track growing-splitting, with propagation and replication of frontal variable F in new nodes 4, 5 and 6, but with remaining of variables H and N in the previous positions 1 and 3;
(d) Echoing from subsequent positions via optimized tracks and continuing support of variables H and N in positions 1 and 3, where not shown subsequent nodes (i.e. following 4, 5, and 6) believably use mobile variable F reaching them; also directly connecting nodes 4 and 1 with removal of non-essential now node 2.

3.7 Conclusions

The Spatial Grasp Model and Technology (SGT) is a paradigm for managing and analyzing distributed systems, particularly those with complex interactions and large scales. It leverages a high-level language called Spatial Grasp Language (SGL) to express operational scenarios that can self-propagate, self-replicate, and self-match within the system. This allows for dynamic control and analysis of distributed environments without relying on central resources. Some key SGT features may be named as follows.

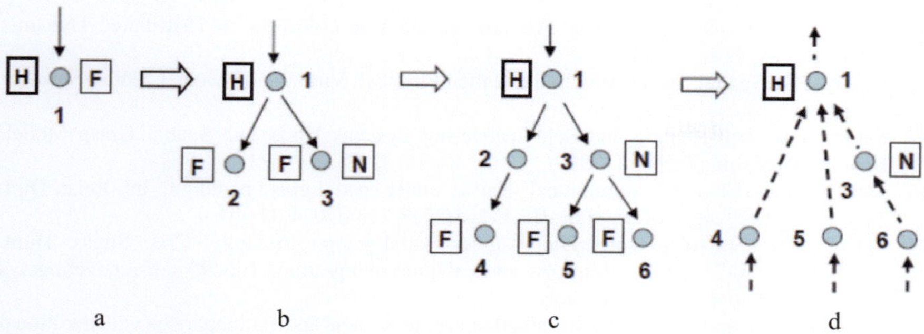

Fig. 3.4 Examples of spatial track operations with the following explanations

Distributed interpretation. SGT utilizes numerous interpreters, often networked, that collectively execute SGL scenarios. These interpreters can be installed in various locations, including human-wearable devices and robotic units, enabling integral teams.

Scenario mobility. SGL scenarios can dynamically move and adapt within the system, covering regions of interest as needed.

Holistic solutions. SGT can provide effective holistic solutions for complex problems that are not achievable by traditional models and technologies which are usually based on the parts to whole principles.

Self-organizing systems. The technology enables the creation of self-organizing systems which can self-evolve, self-analyze, self-modify, and self-recover without central management or control.

References

1. Sapaty, P.S.: A distributed processing system. European Patent N 0389655, Publ. 10.11.93. European Patent Office
2. Sapaty, P.S.: Mobile Processing in Distributed and Open Environments. Wiley, New York (1999)
3. Sapaty, P.S.: Ruling Distributed Dynamic Worlds. Wiley, New York (2005)
4. Sapaty, P.S.: Managing Distributed Dynamic Systems with Spatial Grasp Technology. Springer (2017)
5. Sapaty, P.S.: Holistic Analysis and Management of Distributed Social Systems. Springer (2018)
6. Sapaty, P.S.: Complexity in International Security: A Holistic Spatial Approach. Emerald Publishing (2019)
7. Sapaty, P.S.: Symbiosis of Real and Simulated Worlds under Spatial Grasp Technology. Springer (2021)
8. Sapaty, P.S.: Spatial Grasp as a Model for Space-based Control and Management Systems. CRC Press (2022)
9. Sapaty, P.S.: The Spatial Grasp Model: Applications and Investigations of Distributed Dynamic Worlds. Emerald Publishing (2023)

10. Sapaty, P.S.: Providing Integrity, Awareness, and Consciousness in Distributed Dynamic Systems. CRC Press (2024)
11. Sapaty, P.S.: Spatial Networking in the United Physical, Virtual, and Mental World. Springer (2024)
12. Sapaty, P.S.: Self-Healing and Self-Recovering Systems Under the Spatial Grasp Model. Emerald Publishing Limited (2025)
13. Sapaty, P.S.: Global and international security under spatial grasp paradigm. Int. Relat. Dipl. **12**(2), 72–85 (2024). https://doi.org/10.17265/2328-2134/2024.02.004
14. Sapaty, P.S.: Self-recovering-systems-under-spatial-grasp-technology. Cris. Stress. Hum. Resil. Int. J. **6**(2) (2024). https://www.crisisjournal.org/article/126490-self-recovering-systems-under-spatial-grasp-technology
15. Sapaty, P.S.: Spatial grasp model for distributed management and its comparison with traditional algorithms. Int. Relat. Dipl. **13**(3) (2025). https://doi.org/10.17265/2328-2134/2025.03.004. https://www.davidpublisher.com/Public/uploads/Contribute/689ada551780e.pdf
16. Bondarenko, A.T., Mikhalevich, S.B., Nikitin, A.I., Sapaty, P.S.: Software of BESM-6 computer for communication with peripheral computers via telephone channels. In: Computer Software, vol. 5. Institute of Cybernetics Press, Kiev (1970)
17. Sapaty, P.S.: A method of organization of an intercomputer dialogue in the radial computer systems. In: The Design of Software and Hardware for Automatic Control Systems. Institute of Cybernetics Press, Kiev (1973)
18. Sapaty, S.: A wave language for parallel processing of semantic networks. Comput. Artif. Intell. **5**(4), 289–314 (1986)
19. Sapaty, P.S.: Distributed modeling of cooperative behavior by mobile agents. In: Proceedings of Sixth Conference on Computer Generated Forces and Behavioral Representation, pp. 599–613. IST UCF, Orlando, FL, July 1996
20. Sapaty, P.S.: Distributed artificial brain for collectively behaving mobile robots. In: Proceedings of Symposium and Exhibition Unmanned Systems 2001, Baltimore, MD, July 31–Aug 2, 2001
21. Sapaty, P.S.: Crisis management with distributed processing technology. Int. Trans. Syst. Sci. Appl. **1**(1), 81–92 (2006). ISSN 1751-1461
22. Sapaty, P.S.: Distributed technology for global dominance. In: Proceedings of International Conference Defense Transformation and Net-Centric Systems 2008, as part of the SPIE Defense and Security Symposium. World Center Marriott Resort and Convention Center, Orlando, FL, USA, 16–20 Mar 2008 (Also: Proceedings of SPIE, vol. 6981. In: Suresh R (ed.) Defense Transformation and Net-Centric Systems 2008, 69810T, 3 Apr 2008)
23. Sapaty, P.S.: Distributed technology for global dominance. Keynote lecture. In: Proceedings of the Fifth International Conference in Control, Automation and Robotics ICINCO 2008, The Conference Proceedings, Funchal, Madeira, Portugal, 11–15 May 2008
24. Sapaty, P.S.: Human-robotic teaming: a compromised solution. In: AUVSI's Unmanned Systems North America 2008, San Diego, USA, 10–12 June 2008
25. Sapaty, P.S.: High-level communication protocol for dynamically networked battlefields. In: Proceedings of International Conference Tactical Communications 2009 (Situational Awareness and Operational Effectiveness in the Last Tactical Mile), One Whitehall Place, Whitehall Suite and Reception, London, UK (2009)
26. Sapaty, P.S.: Distributed capability for battlespace dominance. In: Electronic Warfare 2009 Conference and Exhibition. Novotel London West Hotel and Conference Center, London, 14–15 May 2009
27. Sapaty, P.S.: Providing spatial integrity for distributed unmanned systems. In: Proceedings of 6th International Conference in Control, Automation and Robotics ICINCO 2009, Milan, Italy (2009)
28. Sapaty, P.S.: Distributed technology for global control. Book chapter, Lecture Notes in Electrical Engineering, Informatics in Control, Automation and Robotics, vol. 37. Springer, Berlin (2009)
29. Sapaty, P.S.: Gestalt-based integrity of distributed networked systems. In: SPIE Europe Security + Defence. bcc Berliner Congress Centre, Berlin, Germany (2009)

30. Sapaty, P.S.: Remote control of open groups of remote sensors. In: Proceedings of SPIE Europe Security + Defence, Berlin, Germany (2009)
31. Sapaty, P.S.: Tactical communications in advanced systems for asymmetric operations. In: Proceedings of Tactical Communications 2010, CCT Venues, Canary Wharf, London, UK, 28–30 Apr 2010
32. Sapaty, P.S.: High-level technology to manage distributed robotized systems. In: Proceedings of Military Robotics 2010, London, UK, 25–27 May 2010
33. Sapaty, P.S.: Emerging asymmetric threats, Q&A session. In: Tactical Communications 2010, CCT Venues, Canary Wharf, London, UK, 28–30 Apr 2010
34. Sapaty, P.S.: High-level organization and management of directed energy systems. In: Proceedings of Directed Energy Weapons 2010, CCT, Canary Wharf, London, UK, 25–26 Mar 2010
35. Sapaty, P.S.: Unified transition to robotized armies with spatial grasp technology. In: International Summit Military Robotics, London, United Kingdom, 12–13 Nov 2012
36. Sapaty, P.S.: Distributed air and missile defense with spatial grasp technology. Intell. Control Autom. Sci. Res. **3**(2), 117–131 (2012)
37. Sapaty, P.S.: Global electronic dominance. In: 12th International Fighter Symposium, 6th–8th Nov 2012, Grand Connaught Rooms, London, UK
38. Sapaty, P.S.: Providing global awareness in distributed dynamic environments. In: International Summit ISR, London, 16–18 Apr 2013
39. Sapaty, P.S.: Ruling distributed dynamic worlds with spatial grasp technology. Tutorial at the International Science and Information Conference 2013 (SAI), London, UK, 7–9 Oct 2013
40. Sapaty, P.S.: Night vision under advanced spatial intelligence: a key to battlefield dominance. In: International Summit Night Vision 2013, London, 4–6 June 2013
41. Sapaty, P.S.: Integration of ISR with advanced command and control for critical mission applications. In: SMi's ISR Conference, Holiday Inn Regents Park, London, 7–8 Apr 2014
42. Sapaty, P.S.: Unified transition to cooperative unmanned systems under spatial grasp paradigm. Int. J. Trans. Netw. Commun. **2**(2) (2014)
43. Sapaty, P.S.: From manned to smart unmanned systems: a unified transition. In: SMi's Military Robotics, Holiday Inn Regents Park, London, 21–22 May 2014
44. Sapaty, P.S.: Unified transition to cooperative unmanned systems under spatial grasp paradigm. In: 19th International Command and Control Research and Technology Symposium, Alexandria, Virginia, 16–19 June 2014
45. Sapaty, P.S.: Distributed human terrain operations for solving national and international problems. Int. Relat. Dipl. **2**(9) (2014)
46. Sapaty, P.S.: Providing over-operability of advanced ISR systems by a high-level networking technology. In: SMI's Airborne ISR, Holiday Inn Kensington Forum, London, United Kingdom, 26–27 Oct 2015
47. Sapaty, P.S.: Distributed missile defence with spatial grasp technology. In: SMi's Military Space, Holiday Inn Regents Park, London, 4–5 Mar 2015
48. Sapaty, P.S.: Military robotics: latest trends and spatial grasp solutions. Int. J. Adv. Res. Artif. Intell. **4**(4) (2015)
49. Sapaty, P.S.: A brief introduction to the spatial grasp language (SGL). J. Comput. Sci. Syst. Biol. **09**(02) (2015). https://www.researchgate.net/publication/307744279_A_Brief_Introduction_to_the_Spatial_Grasp_Language_SGL
50. Sapaty, P.S.: Organization of advanced ISR systems by high-level networking technology. Math. Mach. Syst. (1) (2016)
51. Sapaty, P.S.: Towards massively robotized systems under spatial grasp technology. J. Comput. Sci. Syst. Biol. **9**(1) (2016)
52. Sapaty, P.S.: Towards wholeness and integrity of distributed dynamic systems. J. Comput. Sci. Syst. Biol. **9**(3) (2016)
53. Sapaty, P.S.: Towards global goal orientation, robustness and integrity of distributed dynamic systems. J. Int. Relat. Dipl. **4**(6) (2016)

54. Sapaty, P.: Spatial grasp language (SGL). In: Advances in Image and Video Processing, vol. 4, no. 1 (2016). https://journals.scholarpublishing.org/index.php/AIVP/article/view/1922/1052
55. Sapaty, P.S.: Mosaic warfare: from philosohpy to model to solution. Mathematical Mach. Syst. (3) (2019)
56. Sapaty, P.S.: Global network management under spatial grasp paradigm. Int. Robot. Autom. J. **6**(3), 134–148 (2020). https://medcraveonline.com/IRATJ/IRATJ-06-00212.pdf
57. Sapaty, P.S.: Global network management under spatial grasp paradigm. Glob. J. Res. Eng. J. Gen. Eng. **20**(5), 58–81. Version 1.0 (2020). https://globaljournals.org/GJRE_Volume20/6-Global-Network-Management.pdf
58. Sapaty, P.: Advanced terrestrial and celestial missions under spatial grasp technology. Aeronaut. Aerosp. Open Access J. **4**(3) (2020). https://medcraveonline.com/AAOAJ/AAOAJ-04-00110.pdf
59. Sapaty, P.: Spatial management of distributed social systems. J. Comput. Sci. Res. **02**(03) (2020). https://ojs.bilpublishing.com/index.php/jcsr/article/view/2077/pdf
60. Sapaty, P.: Towards global nanosystems under high-level networking technology. Acta Sci. Comput. Sci. **2**(8) (2020). https://www.actascientific.com/ASCS/pdf/ASCS-02-0051.pdf
61. Sapaty, P.: Symbiosis of distributed simulation and control under spatial grasp technology. SSRG Int. J. Mob. Comput. Appl. (IJMCA) **7**(2) (2020). http://www.internationaljournalssrg.org/IJMCA/2020/Volume7-Issue2/IJMCA-V7I2P101.pdf
62. Sapaty, P.S.: Symbiosis of real and simulated worlds under global awareness and consciousness. In: Abstract, The Science of Consciousness (TSC) Conferences, Tucson, Arizona (2020). https://eagle.sbs.arizona.edu/sc/report_poster_detail.php?abs=3696
63. Sapaty, P.S.: Spatial grasp as a model for space-based control and management systems. Math. Mach. Syst. (1), 135–138 (2021). http://www.immsp.kiev.ua/publications/articles/2021/2021_1/Sapaty_book_1_2021.pdf
64. Sapaty, P.S.: Managing multiple satellite architectures by spatial grasp technology. Math. Mach. Syst. (1), 3–16 (2021). http://www.immsp.kiev.ua/publications/eng/2021_1/
65. Sapaty, P.S.: Spatial management of large constellations of small satellites. Math. Mach. Syst. (2) (2021). http://www.immsp.kiev.ua/publications/articles/2021/2021_2/02_21_Sapaty.pdf
66. Sapaty, P.S.: Global management of space debris removal under spatial grasp technology. Acta Sci. Comput. Sci. **3**(7) (2021). https://www.actascientific.com/ASCS/pdf/ASCS-03-0135.pdf
67. Sapaty, P.S.: Space debris removal under spatial grasp technology. Netw. Commun. Technol. **6**(1) (2021). https://www.ccsenet.org/journal/index.php/nct/article/view/0/45486
68. Sapaty, P.S.: Relation of spatial grasp paradigm to higher psychological and mental concepts. Acta Sci. Comput. Sci. **4**(12) (2022). https://actascientific.com/ASCS/pdf/ASCS-04-0359.pdf
69. Sapaty, P.S.: Seeing and managing distributed worlds with spatial grasp paradigm. Acta Sci. Comput. Sci. **4**(12) (2022). https://actascientific.com/ASCS/pdf/ASCS-04-0365.pdf
70. Sapaty, P.S.: Comprehending distributed worlds with the spatial grasp paradigm. Math. Mach. Syst. (1) (2022). http://www.immsp.kiev.ua/publications/articles/2022/2022_1/01_22_Sapaty.pdf
71. Sapaty, P.S.: Providing global awareness in distributed dynamic systems. Int. Relat. Dipl. **11**(2), 87–100 (2023). https://doi.org/10.17265/2328-2134/2023.02.002. https://www.davidpublisher.com/Public/uploads/Contribute/6486c3d05a6cc.pdf
72. Sapaty, P.S.: Simulating distributed consciousness with spatial grasp model. Math. Mach. Syst. (3) (2023)
73. Sapaty, P.S.: Network centrality operations under spatial grasp technology. J. Adv. Artif. Intell. Mach. Learn. **1**(1), 1–11 (2023). https://www.scivisionpub.com/journals/articleinpress-journal-of-advances-in-artificial-intelligence-and-machine-learning
74. Sapaty, P.S.: Managing distributed systems with spatial grasp patterns. Math. Mach. Syst. (4) (2023). http://www.immsp.kiev.ua/publications/articles/2023/2023_4/04_23_Sapaty.pdf
75. Sapaty, P.S.: Spatial management of air and missile defence operations. Math. Mach. Syst. (1) (2023). http://www.immsp.kiev.ua/publications/articles/2023/2023_1/01_23_Sapaty.pdf
76. Sapaty, P.S.: Providing distributed system integrity under spatial grasp technology. Math. Mach. Syst. (2) (2023). http://www.immsp.kiev.ua/publications/articles/2023/2023_2/02_23_Sapaty.pdf

77. Sapaty, P.S.: Spatial networks as models for organoid cultures and brain research. Math. Mach. Syst. (1) (2024). ISSN 1028-9763. http://www.immsp.kiev.ua/publications/articles/2024/2024_1/01_24_Sapaty.pdf
78. Sapaty, P.S.: Global and international security under spatial grasp paradigm. Int. Relat. Dipl. **12**(2), 72–85 (2024). https://doi.org/10.17265/2328-2134/2024.02.004. https://www.davidpublisher.com/index.php/Home/Article/index?id=50834.html
79. Sapaty, P.S.: Networking solutions in combined distributed worlds. In: Proceedings of the 6th International Scientific Conference «European Research Materials», pp. 256–275, 30–31 May 2024, Amsterdam, Netherlands (2024). ISBN 978-1-6484-8208-3. https://doi.org/10.5281/zenodo.11432625. https://ojs.publisher.agency/index.php/ERM/issue/view/85/210
80. Sapaty, P.S.: Self-recovering infrastructures and networks under spatial grasp paradigm. In: Proceedings of the 7th International Scientific Conference «Theoretical Hypotheses and Empirical results», pp. 96–112, 11–12 July 2024, Oslo, Norway (2024). ISBN 978-0-2324-6742-0. https://doi.org/10.5281/zenodo.12741275. https://ojs.publisher.agency/index.php/THIR/issue/view/91/222
81. Sapaty, P.S.: Self-recovering systems under spatial grasp technology. Cris. Stress. Hum. Resil. Int. J. (2024) (in print). E-ISSN:2836-1709. https://www.crisisjournal.org/about

Chapter 4
Spatial Grasp Language (SGL)

Abstract The Chapter first reviews existing distributed programming languages including Java, Python, Go, Erlang, and Scala, which are selected for comparison with Spatial Grasp Language (SGL), the key component of the chosen for this book Spatial Grasp Technology, as SGL is fundamentally oriented on distributed programming too. After providing summary of the full SGL syntax, it describes in detail the full repertoire of SGL components and their usage, which include constants, variables and universal rules. Then compares SGL with other languages by explaining their similarity and differences, and concludes the Chapter with confirmation of SGL suitability for highly integral and holistic solutions in any distributed spaces.

4.1 Introduction

Distributed programming languages are those designed to facilitate the development of applications that operate across multiple, interconnected computational units and deal with spatial data or concepts. This combination addresses challenges in areas like geographic information systems (GIS), robotics, and sensor networks, where data is inherently spatial and processing needs to be distributed for performance or resilience.

The Aim of this chapter is to review existing distributed programming languages, provide detailed description of the Spatial Grasp Language serving as key element of the Spatial Grasp technology chosen for this book, and compare SGL with the other languages.

The rest of the chapter reviews existing distributed programming languages including Java, Python, Go, Erlang, and Scala; provides SGL summary with its full syntax; describes SGL constants including information, physical matter, special and custom constants; reports SGL variables of types global, heritable, frontal, nodal, and environmental; presents SGL rules serving as type, usage, movement, creation, echoing, verification, assignment, advancement, branching, transference, exchange, timing, and qualifying; compares SGL with other languages; concludes the chapter and cites used sources.

P. S. Sapaty, *Distributed Management with Spatial Grasp Model*, Studies in Systems, Decision and Control 666, https://doi.org/10.1007/978-3-032-19378-0_4

4.2 Review of Distributed Programming Languages

Distributed programming languages [1] are designed to facilitate the development of software applications that operate across multiple interconnected computers, often referred to as a distributed system. These languages provide features and abstractions that simplify the complexities inherent in such systems, including concurrency, communication, fault tolerance, and scalability. Key characteristics of distributed programming languages include:

Concurrency models. Many distributed languages offer built-in mechanisms for managing concurrent operations, such as threads, goroutines, or actors, to enable parallel execution and efficient resource utilization across multiple nodes.

Communication mechanisms. They provide ways for different parts of a distributed application to exchange information, often through message passing or remote procedure calls.

Fault tolerance. Features handling failures in a distributed environment which are crucial. This can include error handling, recovery mechanisms, and the ability to continue operating even if some nodes fail.

Scalability. Languages and frameworks supporting distributed programming often provide tools to scale applications by adding more nodes to handle increased load or data.

The analyzed for this book distributed programming languages with their main features are as follows.

Java is a high-level, class-based, object-oriented programming language that is designed to have as few implementation dependencies as possible. It is a general-purpose programming language intended to let programmers write once, run anywhere, meaning that compiled Java code can run on all platforms that support Java without the need to recompile [2, 3].

Python is a high-level, general-purpose programming language. Its design philosophy emphasizes code readability with the use of significant indentation. Python is dynamically typed and garbage-collected. It supports multiple programming paradigms, including structured, object-oriented and functional programming. It is often described as a "batteries included" language due to its comprehensive standard library [4, 5].

Go is a statically typed, compiled high-level programming language designed at Google. It is syntactically similar to C, but also has memory safety, garbage collection, structural typing, and CSP-style concurrency. It is often referred to as Golang because of its former domain name, golang.org, but its proper name is Go [6, 7].

Erlang is a general-purpose, concurrent, functional high-level programming language, and a garbage-collected runtime system. The term Erlang is used interchangeably with Erlang/OTP, or Open Telecom Platform (OTP), which consists of the

Erlang runtime system, several ready-to-use components mainly written in Erlang, and a set of design principles for Erlang programs [8, 9].

Scala is a strong statically typed high-level general-purpose programming language that supports both object-oriented programming and functional programming. Designed to be concise, many of Scala's design decisions are intended to address criticisms of Java. Scala source code can be compiled to Java bytecode and run on a Java virtual machine (JVM) [10, 11].

4.3 Spatial Grasp Language (SGL) Summary

Using here and in the rest of this Chapter existing publications related to SGT and SGL, including patent [12], books [13–23], latest papers [24–26], also (far from complete) history [27–92].

In the language summary below, syntactic categories are shown in italics, vertical bar separates alternatives, parts in braces indicate zero or more repetitions with a delimiter at the right, and constructs in brackets are optional. The remaining characters and words are the language symbols (including boldfaced braces).

grasp → *constant* | *variable* | [*rule*] [({ *grasp*,})]
constant → *information* | *matter* | *special*
information → *string* | *scenario* | *number*
string → '{*character*}'
scenario → **{**{*character*}**}**
number → [*sign*]{*digit*}[.{*digit*}[*e*[*sign*]{*digit*}]]
matter → "{*character*}"
special → thru | done | fail | fatal | infinite | nil | any | all |
other | allother | current | passed | existing |
neighbors | direct | forward | backward | neutral | synchronous |
asynchronous | virtual | physical | executive |
engaged | vacant | firstcome | unique
variable → *global* | *heritable* | *frontal* | *nodal* | *environmental*
global → G{*alphameric*}
heritable → H{*alphameric*}
frontal → F{*alphameric*}
nodal → N{*alphameric*}
environmental → TYPE | NAME | CONTENT | ADDRESS | QUALITIES | WHERE|
BACK | PREVIOUS | PREDECESSOR | DOER | RESOURCES | LINK|
DIRECTION | WHEN | TIME | STATE | VALUE | IDENTITY | IN|
OUT | STATUS | SIZE | WEIGHT | LENGTH
rule → *movement* | *creation* | *advancement* | *branching* |
echoing | *verification* | *assignment* | *transference* |
exchange | *timing* | *qualifying* | *type* | *usage*

movement → hop | hopfirst | hopforth | move | shift | follow
creation → create | linkup | delete | unlink
advancement → advance | slide | repeat | align | fringe
branching → branch | sequence | parallel | if | or | and | orsequence |
orparallel | andsequence | andparallel | choose |
quickest | split | cycle | loop | sling | whirl replicate
echoing → state | rake | order | unit | unique | sum | count | first |
last | min | max | random | average | sortup | sortdown |
reverse | element | position | fromto | add | subtract |
multiply | divide | degree | separate | unite | attach |
append | common | withdraw | increment | decrement |
access | invert | apply | location | distance
verification → equal | nonequal | less | lessorequal | more | moreorequal |
bigger | smaller | heavier | lighter | longer | shorter | empty |
nonempty | belong | notbelong | intersect | notintersect | yes | no
assignment → assign | assignpeers
transference → run | call
exchange → input | output | send | receive | emit | get
timing → sleep | allowed
qualifying → contain | release | trackless | free | blind | quit |
abort | stay | lift | seize | exit
type → global | heritable | frontal | nodal | environmental |
matter | number | string | scenario | constant
usage → address | coordinate | content | index | time | speed |
name | place | center | range | doer | node | link | unit

The following sections provide detailed explanations of the mentioned above constructs and their possible usage.

4.4 SGL Constants

Constants can be self-identifiable by the way written or defined by special rules embracing them with arbitrary textual representations.

- **Information**

A *string* can be represented as a sequence of characters embraced with opening-closing single quotation marks:

‘{*character*}’

This sequence can contain other single quotes only if appear in opening-closing pairs, with nesting allowed to any depth. If single words representing information not intersect with other language constructs, the quotes can be omitted. Another string representation may be explicit SGL scenario in (boldfaced here) braces:

{{*character*}}

A *number* can be represented similar to traditional programming languages, as follows (with brackets identifying optional parts and braces for repeating characters).

[*sign*]{*digit*}[.{*digit*}[E[*sign*]{*digit*}]]

- **Physical matter**

Physical matter (physical objects including) in the most general way can be reflected by a sequence of characters embraced with opening-closing double quotation marks:

"{*character*}"

- **Special constants**

These include the following.

thru—indicates (or artificially sets up) control state of the scenario in the current world point as an *absolute success* with possibility of further scenario evolution from this point.
done—indicates (or artificially sets up) control state as a successful termination in the current world point, with blocking further scenario development from this point.
fail—indicates (or artificially sets up) scenario control state as failure in the current world point, without possibility of further scenario development from this particular point.
fatal—indicates (or sets up artificially) fatal failure with massive termination of distributed processes in this and other reached points. The influence of this state may be contained by special higher level rules.
infinite—indicates infinitely large value.
nil—indicates no value at all.
any, **all**, **other, allother**—stating that any one, all (the current one including), any other, or all other elements can be considered and used by some rule.
current—refers to the current element (like node) only, for its further consideration or reentering.
passed—informing that the mentioned elements (like world nodes) have already been passed by the current branch to the current point. They may be accessed easier by backwarding via the history-based distributed control than by a global search.
existing—hinting that world nodes with the given names already exist and should not be created again.
neighbours—stating that the nodes to be accessed are among direct neighbours of the current node, i.e. within a single hop via existing links.
direct—stating that the mentioned nodes should be accessed or created from the current node directly, without considering semantic links to them (even if such links may already exist).

forward, **backward**, **neutral**—allowing control to move from the current node via existing links along, against, or regardless their orientations (non-oriented links can always be traversed in both directions).
synchronous, **asynchronous**—a modifier setting synchronous or asynchronous mode of operations.
virtual, **physical**, **executive**—indicating or setting the type of a node the scenario is currently dealing with (the node can also be of a combined type).
engaged, **vacant**—indicating or setting the state of a resource the current scenario is dealing with.
firstcome—allows the current scenario with its unique identity to enter the world nodes only first time (the capability based on internal node marking mechanisms).
unique—allows the return of only unique elements received from the embraced scenario final positions while omitting duplicates.

- **Custom constants**

Other self-identifiable, or *custom*, constants can be incorporated to be directly processed by the updated SGL if do not conflict with the language syntax, otherwise should be declared by special rules.

4.5 SGL Variables

Their types can be defined by how their names are written, but in general, variables may have any names if their types are declared by special rules, as in Sect. 4.5.

- **Global variables**

This is the most expensive type of SGL variables, with their names starting with capital G and followed by sequences of alphabetic letters and/or digits:

G{*alphameric*}

These variables can exist only in single copies with particular names, being common for both read and write operations to all processes of the same scenario.

- **Heritable variables**

The names of these variables should start with capital H if not defined by a special rule:

H{*alphameric*}

Heritable variables, being created by first assignment to them at some scenario development stage, are becoming common for read–write operations for all subsequent scenario steps (generally multiple, parallel and distributed) evolving from this particular point and wherever is space they may happen to be. This means that such variables are unique only within concrete hereditary scenario developments, to their full depth.

- **Frontal variables**

These are mobile type variables with names starting with capital F, which are propagating in distributed spaces while keeping their contents on the forefronts of evolving scenarios:

F{*alphameric*}

Each of these variables is serving only the current scenario branch operating in the current world point. They cannot be shared with other branches evolving in the same or other world points, while always accompanying the scenario control. If the scenario splits into individual branches, these variables are replicated with the same names and contents and serve these branches independently.

- **Nodal variables**

Variables of this type, their identifiers starting with capital N, are a temporary and *exclusive property of the world points* visited by SGL scenarios, which can create, change or remove them:

N{*alphameric*}

Capable of being shared by all scenario branches visiting these nodes, they are created by first assignment to them and stay in the node until removed explicitly, or at least a single scenario that passed through this node still remains active. These variables also cease to exist when nodes they associate with are removed by any scenario reaching them.

- **Environmental variables**

These are special variables with reserved names (all in capitals) which allow us to have access to physical, virtual, and execution worlds when they are navigated by SGL scenarios, also to some parameters of the language interpretation system itself.

TYPE—indicates the type of a node the current scenario step associates with and returns a verbal expression of the node's type (i.e. virtual, physical, executive or their combination). It can also change the existing node's type by assigning to it another value (simple or combined).
NAME—returns name of the current node as a string of characters (only if the node has virtual or executive dimension or both). Assigning to this variable when staying in the node can change the node's name.
CONTENT—returns content of the current node (if it has virtual or executive dimension or both) as any text, vector, or nested structure of multiple texts, etc. Assigning to this variable when staying in the node can set up or change the node's content.

ADDRESS—returns a unique address of the node having virtual dimension. This is read-only variable as node addresses are set up automatically by the underlying distributed SGL interpretation system during node's creation, or by an external system (like internet address of the node).
QUALITIES—identifies a list of selected formalized physical parameters associated with the current physical position, or node, depending on the chosen implementation and application (for example, these may be temperature, humidity, air pressure, visibility, radiation, noise or pollution level, density, salinity, etc.). These parameters (generally as a list of values) can be obtained by reading the variable, and some can be changed, thus influencing the world from its particular point.
WHERE—keeps world coordinates of the current physical node (or the one having physical dimension too) in the chosen coordinate system. These coordinates can be obtained by reading this variable. Assigning a new value to this variable (with possible speed added) can cause physical movement into the new position.
BACK—keeps internal system link to the preceding world node (virtual, executive or combined one) allowing the scenario to most efficiently return to the previously occupied node, if needed.
PREVIOUS—refers to the absolute and unique address of the previous virtual node (or combined one with executive and/or physical dimensions), allowing us to return to the node directly. Its content, unlike with variable BACK, can be lifted, recorded, and used elsewhere in the scenario (but not changed).
PREDECESSOR—refers to the name of preceding world node (the one with virtual or executive dimension, visited just before the current one). Its content can be lifted, recorded and subsequently used, including for organization of direct hops to this node (on the most expensive level, however).
DOER—keeps the name of the device (say, laptop, robot, satellite, smart sensor, or a specially equipped human) which interprets the current SGL code in the current world position. This device can be initially chosen for the scenario automatically from the list of recommended devices or just picked up from those expected available. It can also be appointed explicitly by assigning its name to DOER.
RESOURCES—may keep a list of available or recommended resources (human, robotic, electronic, mechanical etc., by their types or names) which can be used for planning and execution of the current and subsequent parts of the SGL scenario. This list can also contain potential doers which may appear (by their names) in variables DOER.
LINK—keeps the name (same as content) of the virtual link which has just been passed. Assigning a new value to it can change the link's content/name. Assigning nil or empty to LINK removes the link passed.
DIRECTION—keeps direction (along, against, or neutral) of the passed virtual link. Assigning to this variable values like plus, minus, or nil (same as +, −, or empty) can change its orientation or make the link non-oriented.
WHEN—assigning value to this variable sets up an absolute starting time for the following scenario branch (i.e. starting with the next operation), thus allowing us to suspend and schedule operations and their groups in time.

TIME—returns the current absolute system time as the read-only global variable.
STATE—can be used for explicit setting resultant control state of the current scenario step by assigning to it one of the following constants: thru, done, fail, or fatal, which will influence further scenario development from the current world point (and in a broader scale in the case of fatal). These control states are also generated implicitly and automatically on the results of success or failure of different operations (belonging to the internal interpretation mechanisms of SGL scenarios). Reading STATE will always return thru as this could be possible only if the previous operation terminated with thru too, thus letting this operation to proceed.
VALUE—when accessed, returns the *resultant value* of the latest, i.e. preceding, operation (say, an assignment to it or any other variable, unassigned result of arithmetic or string operation, or just naming a variable or constant).
IDENTITY—keeps identity, or color, of the current SGL scenario or its branch, which propagates together with the scenario and influences grouping of different nodal variables under this identity at world nodes reached. This allows different scenarios or their branches with personal identities to be protected from influencing each other, even if they are using same named nodal variables in the same world nodes.
IN—special variable requesting and reading data from the outside world in its current point. The received data is becoming the resultant value of the reading operation.
OUT—special variable allowing us to issue information from the scenario in its current point to the outside world, by assigning the output value to this variable.
STATUS—retrieving or setting the status of (especially doer) node in which the scenario is currently staying (like engaged or vacant, possibly, with a numerical estimate of the level of engagement or vacancy). This feedback from implementation layer on the SGL scenario layer can be useful for a higher-level supervision, planning and distribution of resources executing the scenario rather than doing this implicitly and automatically.
SIZE, **WEIGHT**, **LENGTH**—correspondingly, size of the node and weight and length of the link, as additional parameters representing certain real world components.

4.6 SGL Rules

- **Type**

These rules explicitly assign types to different constructs, with their existing repertoire following.

global, **heritable**, **frontal**, **nodal**, **environmental**—allow different types of variables to have any alphanumeric names rather than those oriented on self-identification, as explained before. These names will represent variables with needed types in the subsequent scenario developments unless redefined by these rules too.
matter, **number**, **string**, **scenario**, **constant**—allow arbitrary results obtained by the embraced scenario to properly represent the needed values rather than using self-identifiable representations mentioned before.

- **Usage**

These rules explain how to use the information units they embrace.

address—identifies the embraced value (which may also be arbitrary scenario producing this value) as an address of a virtual node.
coordinate—identifies the embraced value as physical coordinates (say, one, two or three dimensional).
content—identifies the embraced operand as a content (or contents).
index—identifies the embraced operand as an index (or indices) which may represent orders of elements in a list for its search by the index operation.
time—informs that the embraced operand represents time value.
speed—informs that the embraced operand represents a value of speed.
name—identifies the embraced operand as a name (say, of a virtual or executive node or nodes).
center—depending on applications, indicates that virtual address or physical coordinates embraced may relate to the center of some region.
range—identifies virtual or physical distance that can, for example, be used as a threshold for certain operations in distributed spaces, especially those evolving from a chosen or expected center.
doer—identifies the embraced name or any other value as belonging to executive node (like human, robot, server, satellite, smart phone, etc.).
node—(or **nodes**, if more appropriate) identifies the embraced value or values as keeping names of nodes having virtual or/and executive dimensions.
link—(or **links**, when more appropriate) informs that the embraced value or values represent names of links connecting nodes.
unit—using the embraced elements as an integral operand.

- **Movement**

It may result in virtual hopping to the existing nodes (the ones having virtual or/and executive dimensions) or in real movement to new physical locations, subsequently starting the remaining scenario if any (with current frontal variables and control) in the nodes reached.

hop—sets electronic propagation to a node (or nodes) in virtual, execution, or combined spaces (the latter may have physical dimension too), directly or via semantic links connecting them with the starting node. In case of a direct hop, except destination node name or address, special modifier direct may be included into parameters of the rule. If the hop is to take place from a node to a particular node via existing link, both destination node name/address and link name (with orientation if appropriate) should be parameters of the rule.

hopfirst—modification of the hop rule allowing it to come to a node only first time (for the scenario with certain identity), which is based on internal interpretation mechanism marking the nodes visited.

hopforth—modification of the previous rule allowing it to hop to a node which is not the one just visited before, i.e. excluding the return to the previous node. It may be considered as a restricted variant of hopfirst rule.

move—sets real movement in physical world from the current node with physical dimension (which may be combined with virtual and executive ones) to a particular location given by coordinates in a chosen coordinate system. Speed value for the physical propagation by move may be given as an additional parameter.

shift—differs from the move only in that movement in physical world is set by deviations of physical coordinates from the current position rather than by their absolute values.

follow—allows us to move in virtual, physical and combined spaces using already recorded and saved paths from a starting node to the destinations reached, entering the latter again.

- **Creation**

These rules create or remove nodes and/or links leading to them during distributed world navigation. After termination of the rules their resultant values will correspond to the names of reached nodes with states thru in them, and the next scenario steps, if any, may start from all these nodes. If the creation or removal operation fails, its resultant value will be nil and control state fail in the node the rule started.

create—starting in the current world position, creates either new isolated nodes, or new virtual link-node pairs. For the first case, the rule may additionally use modifier "direct" explicitly indicating direct nodes creation. For the second case, the rule is supplied with names and orientations of new links and names of new nodes these links should lead to, which may be multiple.

linkup—restricts the previous rule by creating only links with proper names from the current node to the already existing nodes given by their names or addresses.

delete—starting from the current node, removes links together with nodes they should lead to. Links and nodes to be removed should be either explicitly named or represented by modifiers any or all. Using modifier direct will allow us to remove such node (or nodes) from the current node directly.

unlink—removes only links leading to neighboring nodes where, similar to the previous case, they should be explicitly named or modifiers any or all used instead.

- **Echoing**

This class of rules, oriented on various aspects of data and knowledge processing, contains the following rules which may use local and remote values for different operations. The listed rules use terminal world positions reached by the embraced scenario with their control states and associated final values (which may be local or arbitrarily remote) to obtain the resultant state and value in the location where the rule started.

state—returns the resultant generalized state of the embraced SGL scenario upon its completion, whatever its complexity and space coverage may be. This state being the result of ascending fringe-to-root generalization of terminal states of the scenario embraced, where states with higher power (their sequence from maximum to minimum values as: fatal, thru, done, fail) dominate in this potentially distributed and parallel process.
rake—returns a list of final values of the scenario embraced in arbitrary order, which may, for example, be influenced by the order of completion of branches and times of reaching their final destinations. Additionally using unique as modifier, described before, the rule will result in collecting only unique values.
order—returns an ordered list of final values of the scenario embraced corresponding to the order of launching related branches rather than the order of their completion. For potentially parallel branches, these orders may, for example, relate to how they were activated, possibly, with the use of time stamps upon invocation.
unit—returns a list of values while arranging it as an integral parenthesized unit which should not be mixed with elements returned from other branches which may represent integral units too, to form (potentially hierarchical and nested) lists of lists of the obtained values at higher levels.
sum—returns the sum of all final values of the scenario embraced.
count—returns the number of all resultant values associated with the scenario embraced.
first, **last**, **min**, **max**, **random**, **average**—return, correspondingly, the first, last, minimum, maximum, randomly chosen, or average value from all terminal values returned by the scenario embraced.
sortup, **sortdown**—return an ordered list of values produced by the embraced scenario operand, starting from minimum or maximum value and ending, correspondingly, with maximum or minimum one.
reverse—changes to the opposite the order of values from the embraced operand.
element—returns the value of an element of the list on its left operand requested by index or content given by the right operand. If the right operand is itself a list of indices or contents, the result will be the list of corresponding values from the left operand. If element is used within the left operand of assignment, instead of returning values it will be providing an access to them in order to be updated.
position—returns the index (or indices) of the list on its left operand requested by the content given by the right operand. There may be more than a single index

returned if same content repeats in the list, or if the right operand is itself a list of contents.

fromto—returns an ordered list of digital values by naming its first (operand 1) and last (operand 2) elements as well as step value (operand 3) allowing the next element to be obtained from the previous one. Another modification (depending on implementation) may take into account the starting element, step value, and the number of needed elements in the list.

add, **subtract**, **multiply**, **divide**, **degree**—perform corresponding operations on two or more operands embraced, each potentially represented by arbitrary scenario with local or remote results. If the operands themselves provide multiple values, as lists, these operations are performed between peer elements of these lists, with the resultant value as a list too.

separate—separates the left operand string value by the string at the right operand used as a delimiter in a repeated manner for the left string, with the result being the list of separated substring values. If the right operand is a list of delimiters, its elements will be used sequentially and cyclically unless the string at the left is fully processed/partitioned.

unite—integrates the list of values at the left (as strings, or to be converted into strings automatically) by a repeated delimiter as a string too (or a cyclically used list of them) at the right into a united string.

attach—produces the resultant string by connecting the right string operand directly to the end of the left one. If operands are lists with more than one element, the attachment is made between their peer elements, receiving the resultant list of united strings. This rule can also operate with more than two operands.

append—forms the resultant list from left and right operands by appending the latter to the end of the former as individual elements, where both operands may be lists themselves. More than two operands can be used too for this operation.

common—returns intersection of two or more lists as operands, with the result including only same elements of all lists, if any.

withdraw—its returned result will be the first element of the list provided by the embraced operand, along with withdrawing this element from the head of the list (thus simultaneously changing the content of the variable). This rule can have another operand providing the number of elements to be withdrawn in one step and represented as the result.

increment—adds 1 (one) to the value of the embraced operand which will be the result on this rule, thus simultaneously changing the content of the operand itself. If another value, not 1, to be added, the second operand can keep this value.

decrement—behaves similar to the previous rule increment but subtracts rather than adds 1 from the value of the embraced operand, with the content of the latter simultaneously changed. Second operand can be used if the value to subtract not equals 1.

access—by embracing a scenario or its branch, returns a reference to the internal history-based optimized and recorded structure leading from the rule-activation node to the reached terminal nodes. This reference can be remembered (say, in a variable) and subsequently used from the same starting node for reaching exactly the same terminal nodes again.

invert—changes the sign of a value or orientation of a link to the opposite, while producing no effect on zero values or non-oriented links.

apply—organizes application of the first operand as one or a set of rules described above operating jointly from the same starting point to the same second scenario operand, which may be arbitrary. If multiple application rules engaged on the first operand, the obtained results on the second operand can be multiple too.

location—returns world locations of the final nodes reached by the embraced scenario, which mean for virtual nodes their network addresses, and for physical nodes physical coordinates.

distance—returns distance between two physical points defined by absolute physical coordinates expressed by its parameters, where each one can be represented by an arbitrary scenario.

- **Verification**

These rules provide control state thru or fail reflecting the result of concrete verification procedure, also nil as own resultant value, while remaining after completion in the same world positions where they started,

equal, **nonequal**, **less**, **lessorequal**, **more**, **moreorequal**, **bigger**, **smaller**, **heavier**, **lighter**, **longer**, **shorter**—make corresponding comparison between left and right operands, which can represent information or physical matter/objects, or both. In case of vector operands, state thru appears only if all peer values satisfy the condition set up by the rule (except nonequal, for which even a single non-correspondence between any peers will result in overall thru).

empty, **nonempty**—checks for emptiness (i.e. non-existence of anything, same as nil) or non-emptiness (existence) of the resultant value obtained from the embraced scenario.

belong, **notbelong**—verifies whether the left operand value (single or a list with all its elements) belongs as a whole to the right operand generally represented as a list (which may have a single element too).

intersect, **notintersect**—verifies whether there are common elements (values) in the left and right operands, considered generally as lists. More than two operands can be used for these rules too, with at least a single same element to be present in all of them to result thru for intersect, or no such elements for notintersect.

yes—verifies generalized state of the embraced scenario providing own control state thru in case of thru or done from the entire scenario, and control state fail in case of resultant fail or fatal (thus allowing to continue from the node where the rule started only in case of success of the embraced scenario).

no—verifies generalized state of the embraced scenario resulting with own control state thru in case of fail or fatal from the scenario, and control state fail in case of thru or done (i.e. allowing to continue from the rule's starting node only in case of failure of the embraced scenario).

- **Assignment**

These rules assign the result of the right scenario operand (which may be arbitrarily remote, also represent a list of values which can be nested) to the variable or set of variables directly named or reached by the left scenario operand, which may be remote too. The left operand can also provide pointers to certain elements of the reached variables which should be changed by the assignment rather than the whole contents of variables.

assign—assigns the same value of the right operand (which may be a list of values) to all values (like, say, node names) or variables accessed by the left operand (or their particular elements pointed, which may themselves become lists after assignment, thus extending the lists of contents of these variables). If the right operand is nil or empty, the left operand nodes or variables as a whole (or only their certain elements pointed) will be removed.
assignpeers—assigns values of different elements of the list on the right operand to different values or variables (or their pointed elements) associated with the destinations reached on the left operand, in a peer-to-peer mode.

- **Advancement**

These rules can organize forward or "in depth" advancement in space and time of the embraced scenarios separated by comma. They can evolve within their sequence in synchronous or asynchronous mode using modifiers synchronous or asynchronous (the second one optional, as asynchronous being a default mode).

advance—organizes stepwise scenarios advancement in physical, virtual, executive or combined spaces, also in a pure computational space (the latter when staying in the same world nodes with certain data processing, thus moving in time only). For this, the embraced SGL scenario-operands are used in a sequence, as written, where each new scenario shifts to and applies from all terminal world points reached by the previous scenario. The resultant world positions and values on the whole rule are associated with the final steps of the last scenario on the rule. And the rule's resultant state is a generalization of control states associated with these final steps.
slide—works similar to the previous rule unless a scenario in their sequence fails to produce resultant state thru or done from some world node. In this case the next scenario from the sequence will be applied from the same starting position of the previous, failed, scenario, and so on. The resultant world nodes and values in them will be from the last successfully applied scenarios (not necessarily the same from their sequence, as independently developing in different directions).
repeat—invokes the embraced scenario as many times as possible, with each new iterations taking place in parallel from all final positions with state thru reached by the previous invocations. If some scenario iteration fails, its current starting position with its value will be included into the set of final positions and values on the whole rule (this set may have starting positions from different failed iterations

which developed independently in a distributed space). By supplying additional numeric modifier to this rule, it is possible to explicitly limit the number of allowed scenario repetitions.

align—is based on confirmation of full termination of all activities of the embraced operand-scenario in all its final nodes. Only after this, the remaining scenario part, if any, will be allowed to continue from all the nodes reached.

fringe—allows us to establish certain constraints (say, by additional parameters) on the terminal world nodes reached by the embraced scenario with final values in them, to be considered as starting positions for the following scenario parts. For example, by comparing values in all terminal nodes and allowing the scenario to continue from a node with maximum or minimum value, integrating this rule with previously mentioned rules max or min like **max_fringe** or **min_fringe** can also be possible.

For the advancement rules, frontal variables propagate on the forefronts together with advancement of control and operations in distributed spaces, with next scenarios or their iterations picking up frontal variables brought to their starting points by the previous scenarios, being also replicated if this control automatically splits into different branches.

- **Branching**

These rules allow the embraced set of scenario operands to develop "in breadth", each from the same starting position, with the resultant set of positions and order of their appearance depending on the logic of a concrete branching rule The rest of the SGL scenario will be developing from all or some of the positions and nodes reached on the rule. The branching may be static and explicit if we have a clear set of individual operand scenarios separated by comma. It can also be implicit and dynamic, as explained later. For all branching rules that follow, the frontal variables associated with the rule's starting position will be replicated together with their contents and used independently within different branches.

branch—the most general and neutral variant of branching, with logical independence of the scenario operands from each other and any possible order of their invocation and development from the starting position. The resultant set of positions reached with their associated values will unite all terminal positions and values on all scenario operands involved under branch. The resultant control state on the whole rule will be based on generalization of the generalized control states on all scenario branches.

sequence—organizing strictly sequential invocation of all scenario operands regardless of their success or failure, and launching the next scenario only after full completion of the previous one. The resultant set of positions, values, and rule's global control state will be similar to branch. However, the final results may vary due to different invocation order of the scenario operands and possible common information used.

parallel—organizing fully parallel development of all scenario operands from the same starting position. The resultant set of positions, values, and rule's control state will be similar to the previous two rules, but may not be the same, as explained before.

if—may have three scenario operands. If the first scenario results with generalized termination state thru or done, the second scenario is activated, otherwise the third one will be launched. The resultant set of positions and associated values will be the same as achieved by the second or third scenarios after their completion. If the third operand-scenario is absent and the first one results with fail, or only the first operand is present regardless of its success or failure, the resultant position will be the one the rule started from, with state thru and value it had at the start.

or—allows only one operand scenario with the resulting state thru or done, without any predetermined order of their invocation, to be registered as resultant, with the final positions and associated values on it to be the resulting ones on the whole rule. The activities of all other scenario operands and all results produced by them will be terminated and cancelled.

orsequence—will launch the scenario operands in strictly sequential manner, one after the other as they are written, waiting for their full completion before activating the next operand, unless the first one in the sequence replies with generalized state thru or done (providing the result on the rule as a whole). Invocation of the remaining scenarios in the sequence will be skipped.

orparallel—activates all scenario operands in parallel from the same current position, with the first one in time replying with generalized thru or done being registered as the resultant branch for the rule. All other branches will be forcefully terminated without waiting for their completion (or just ignored).

and—activates all scenario operands from the same position, without any predetermined order, demanding all of them to return generalized states thru or done. If at least a single operand returns generalized fail, the whole rule results with state fail and nil value in the starting position while terminating the development of all other branches which may still be in progress. If all operand scenarios succeed, the resulting set of positions unites all resultant positions on all scenario-operands with their associated values.

andsequence—activates scenario-operands from the same position in the written order, launching next scenario only after the previous one completes with thru or done, and terminating the whole rule when the current scenario results with fail. The remaining scenario operands will be ignored, and all results produced by this and all previous operands will be removed.

andparallel—activates in parallel all scenario operands from the same world position, terminating the rule when the first one in time results with fail, while aborting activity of all other operands and removing all results produced by the rule.

choose—chooses a scenario branch in their sequence before its execution, using additional parameters among which, for example, may be its numerical order in the sequence (or a list of such orders to select more than one branch). This rule can also be aggregated with other rules like first, last or random, by forming combined ones: **choose_first**, **choose_last**, **choose_random**.

quickest—selects the first branch in time replying its complete termination, regardless of its generalized termination state, which may happen to be fail too, even though other branches (to be forcefully terminated now) could respond later with thru or done. The state, set of positions on this selected branch, and their associated values (if any) will be taken as those for the whole rule. This rule assumes that different branches are launched independently and in parallel.

cycle—repeatedly invokes the embraced scenario from the same starting position until its resultant generalized state remains thru or done, where on different invocations the same or different sets of resultant positions (with same or different values) may emerge. The resultant set of positions on the whole rule will be an integration of all positions on all successful scenario invocations with their associated values. The next scenario will be developing from all these world positions reached.

loop—differs from the previous rule in that the resultant set of positions on it being only the set produced by the *last* successful invocation of the embraced scenario.

sling—invokes repeatedly the embraced scenario until it provides state thru or done, always resulting in the same starting position with state thru and its previously associated value.

whirl—endlessly repeating the embraced scenario from the starting position regardless of its success or failure and ignoring any resultant positions or values produced. External forceful termination of this construct may be needed, like using first in time termination of another, competitive, branch.

split—performs, if needed, additional (and deeper than usual) static or dynamic partitioning of the embraced scenario into different branches, especially in complex and not clear at first sight cases, all starting from the same current position. It may be used alone or in combination with the above mentioned branching rules while preparing separate branches for these rules, ahead of their invocation.

replicate—replicates the scenario given by its second operand, providing the number of its copies given by the first operand, with each copy behaving as independent branch starting from the same current world position.

- **Transference**

The rules organize transference of control in distributed scenarios.

run—transfers control to the SGL code treated as a procedure and being a result of invocation of the embraced scenario. The procedure (or a list of them) obtained and activated in such a way can produce a set of world positions with associated values and control states as the result on the rule, similar to other rules.

call—transfers control to the code produced by the embraced scenario which may represent activation of external systems. The resultant world position on call will be the same where the rule started, with value in it of what has been returned from the external call and state thru if the call was successful.

- **Exchange**

These rules include the following.

input—provides input of external information or physical matter (objects) on the initiative of SGL scenario, resulting in the same position but with value received from the outside. The rule may have an additional argument clarifying a particular external source from which the input should take place.
output—outputs the resultant value obtained by the embraced scenario, which can be multiple, with the same resultant position as before and associated value sent outside (in case of physical matter, the resultant value may depend on the applications). The rule may have an additional pointer to a particular external sink.
send—staying in the current position associated with physical, virtual, executive (or combined) node, sends information or matter obtained by the scenario on the first operand to other similar node given by name, address or coordinates provided by the second operand, assuming that a companion rule receive is already engaged there. The rule may have an additional parameter setting acceptable time delay for the consumption of this data at the receiving end.
receive—a companion to rule send, naming the source of data to be received from (defined similarly to the destination node in send). Additional timing (as a second operand) may be set too, after expiration of which the rule will be considered as failed. In case of successful receipt of the data, the rule will result in the same world position and the value obtained.
emit—depending on implementation and technical capabilities, can trigger nonlocal to global continuous broadcasting of the data obtained by the embraced scenario (possibly, with tagging of this source, also setting the emission frequency). Another operand providing time allowed for this broadcasting may be present too. No feedback from possible consumers of the sent data is expected.
get—tries to receive data which can be broadcast from some source (say, identified by its tag or frequency), with resultant value as the received data and state thru. No synchronization with the data emitting node is expected.

- **Timing**

These rules are dealing with conditions related to a time interval for the scenarios they embrace.

sleep—establishes time delay defined by the embraced scenario operand, with suspending activities of this particular scenario branch in the current node. The rule's starting position and its existing value, also state thru, will be the result on the rule after the time passed.

allowed—sets time limit by the first operand for an activity of the scenario on the second operand. If the scenario terminates before expiration of this time frame, its resultant positions with values and states will define the result on this rule. Otherwise the scenario will be forcibly aborted.

- **Qualifying**

These rules are providing certain qualities or abilities, also setting constraints or restrictions to the scenarios they embrace, as follows.

contain—restricts the spread of abortive consequences caused by control state fatal within the ruled scenario. This state may appear automatically and accidentally in different scenario development points or can be assigned explicitly to environmental variable STATE, triggering emergent completion of multiple scenario processes and removal of temporary data associated with them. The resultant position on the rule contain having state fatal inside its scenario will be the one it started from, with value nil and state fail. If fatal did not appear inside this rule, its result will be defined by the completion of the embraced scenario.
release—allows the embraced scenario develop free from the main scenario, abandoning bilateral control links with it (the main scenario after the rule's activation "will not see" this construct any more). The released, now independent, scenario will develop as a whole in the usual way, using its standard subordination and control mechanisms.
trackless—allows the whole embraced scenario to develop absolutely free from any previous stages (i.e. without saving any control and information links with them) like a real virus, being an extreme and global variant of release. Under this rule, frontal variables carrying information between different stages will work as usual.
free—differs from the previous case in that despite its independence and control freedom from the main scenario, as before, the embraced scenario will nevertheless be obliged to return the final data obtained in its terminal positions to the main scenario.
blind, quit, abort—after full completion of the embraced scenario, these rules result in the same position the rule started with respective states done, fail, or fatal, thus preventing further scenario development from this point (also triggering nonlocal termination and cancellation processes in case of fatal).
stay—whatever the scenario embraced and its evolution in space, the resultant position will always be the same this rule started from, with value nil and state thru in it. If the ruled scenario is omitted, this rule standing alone just represents an empty operation in the current point, which can also be used for declaring a new scenario branch.
lift—lifts blocking of the further scenario developments set up by states done in the embraced scenario wherever it happened to emerge, substituting them with thru and allowing further developments from all such positions.

seize—establishes, or seizes, an absolute control over the resources associated with the current virtual, physical, executive or combined node, blocking these from any other accesses and allowing only the embraced scenario to work with them (thus preventing possible competition for the node's assets which may lead to unexpected results). This resource blockage is automatically lifted after the embraced scenario terminates. If more than two scenarios are competing for the node's resources, they will be organized in a FIFO manner at the node.
exit—terminates the innermost loop in which it is included.

4.7 Comparison of SGL with Other Languages

In comparison between the mentioned in Sect. 4.2 most popular distributed programming languages [1–11] and SGL we can conclude the following.

> SGL may also be considered as a programming language for distributed systems, which can involve multiple interconnected computers working together to achieve a common goal.
> It is also object-oriented and network-centric language capable of coding everything from mobile apps to big data applications, and can be used as a platform in itself.
> SGL can used in different fields such as AI, data science, web development, defense, security, sociology, etc., also allowing for unlimited concurrency.
> SGL is oriented on code readability with the use of indentation, especially due to its highly recursive nature and unlimited code hierarchy and compactness.
> SGL can support multiple programming paradigms, including structured, object-oriented and functional programming, and its high-level and clear code helps programmers get started programming right away.
> SGL high-level scenarios are very concise, usually much shorter than in other languages (up to hundred times than in C or Java).
> SGL can run anywhere and on any platform via its hardware or software interpreter (as was written in C).
> SGL also includes embedded garbage collection during runtime, where the used parts of the scenarios and obtained data not needed any more are automatically self-deleted.
> Despite widely used for distributed programming too, SGL is fundamentally based on quite different principles allowing for ***highly integral holistic solutions in wavelike spatial coverage and parallel pattern matching*** mode, rather than representing solutions as communicating parts or agents.

4.8 Conclusions

After reviewing the existing popular distributed programming languages, we provided full description of Spatial Grasp Language (SGL) with its syntax and main constructs, also compared main features of SGL with the other languages. Despite capability of SGL for being used similarly to other distributed programming languages, it is however fundamentally based on quite different and more advanced principles, allowing for ***highly integral holistic solutions*** in any terrestrial and celestial spaces. SGL philosophy and constructs allow SGL programmer, more correctly, ***spatial composer and ruler***, to have a feeling of ***direct existence*** in the distributed spaces, while observing, managing, and modifying them as needed.

References

1. Programming languages for distributed systems, 03 July 2024. https://www.geeksforgeeks.org/programming-languages-for-distributed-systems/
2. Java (programming language). https://en.wikipedia.org/wiki/Java_(programming_language)
3. What is Java? https://aws.amazon.com/what-is/java/
4. Python (programming language). https://en.wikipedia.org/wiki/Python_(programming_language)
5. Master Python programming. https://www.programiz.com/python-programming
6. Go (programming language). https://en.wikipedia.org/wiki/Go_(programming_language)
7. Donovan, A., Kernighan, B.: Go Programming Language. Addison-Wesley Professional Computing
8. Erlang (programming language). https://en.wikipedia.org/wiki/Erlang_(programming_language)
9. Erlang tutorial. https://www.tutorialspoint.com/erlang/index.htm
10. Scala (programming language). https://en.wikipedia.org/wiki/Scala_(programming_language)
11. Scala programming language, 11 Dec 2023. https://www.geeksforgeeks.org/scala-programming-language/
12. Sapaty, P.S.: A distributed processing system. European Patent N 0389655, Publ. 10.11.93. European Patent Office
13. Sapaty, P.S.: Mobile Processing in Distributed and Open Environments. Wiley, New York (1999)
14. Sapaty, P.S.: Ruling Distributed Dynamic Worlds. Wiley, New York (2005)
15. Sapaty, P.S.: Managing Distributed Dynamic Systems with Spatial Grasp Technology. Springer (2017)
16. Sapaty, P.S.: Holistic Analysis and Management of Distributed Social Systems. Springer (2018)
17. Sapaty, P.S.: Complexity in International Security: A Holistic Spatial Approach. Emerald Publishing (2019)
18. Sapaty, P.S.: Symbiosis of Real and Simulated Worlds under Spatial Grasp Technology. Springer (2021)
19. Sapaty, P.S.: Spatial Grasp as a Model for Space-based Control and Management Systems. CRC Press (2022)
20. Sapaty, P.S.: The Spatial Grasp Model: Applications and Investigations of Distributed Dynamic Worlds. Emerald Publishing (2023)
21. Sapaty, P.S.: Providing Integrity, Awareness, and Consciousness in Distributed Dynamic Systems. CRC Press (2024)

22. Sapaty, P.S.: Spatial Networking in the United Physical, Virtual, and Mental World. Springer (2024)
23. Sapaty, P.S.: Self-Healing and Self-Recovering Systems Under the Spatial Grasp Model. Emerald Publishing Limited (2025)
24. Sapaty, P.S.: Global and international security under spatial grasp paradigm. Int. Relat. Dipl. **12**(2), 72–85 (2024). https://doi.org/10.17265/2328-2134/2024.02.004
25. Sapaty, P.S.: Self-recovering-systems-under-spatial-grasp-technology. Cris. Stress. Hum. Resil. Int. J. **6**(2) (2024). https://www.crisisjournal.org/article/126490-self-recovering-systems-under-spatial-grasp-technology
26. Sapaty, P.S.: Spatial grasp model for distributed management and its comparison with traditional algorithms. Int. Relat. Dipl. **13**(3) (2025). https://doi.org/10.17265/2328-2134/2025.03.004. https://www.davidpublisher.com/Public/uploads/Contribute/689ada551780e.pdf
27. Bondarenko, A.T., Mikhalevich, S.B., Nikitin, A.I., Sapaty, P.S.: Software of BESM-6 computer for communication with peripheral computers via telephone channels. In: Computer Software, vol. 5. Institute of Cybernetics Press, Kiev (1970)
28. Sapaty, P.S.: A method of organization of an intercomputer dialogue in the radial computer systems. In: The Design of Software and Hardware for Automatic Control Systems. Institute of Cybernetics Press, Kiev (1973)
29. Sapaty, S.: A wave language for parallel processing of semantic networks. Comput. Artif. Intell. **5**(4), 289–314 (1986)
30. Sapaty, P.S.: Distributed modeling of cooperative behavior by mobile agents. In: Proceedings of Sixth Conference on Computer Generated Forces and Behavioral Representation, pp. 599–613. IST UCF, Orlando, FL, July 1996
31. Sapaty, P.S.: Distributed artificial brain for collectively behaving mobile robots. In: Proceedings of Symposium and Exhibition Unmanned Systems 2001, Baltimore, MD, July 31–Aug 2, 2001
32. Sapaty, P.S.: Crisis management with distributed processing technology. Int. Trans. Syst. Sci. Appl. **1**(1), 81–92 (2006). ISSN 1751-1461
33. Sapaty, P.S.: Distributed technology for global dominance. In: Proceedings of International Conference Defense Transformation and Net-Centric Systems 2008, as part of the SPIE Defense and Security Symposium. World Center Marriott Resort and Convention Center, Orlando, FL, USA, 16–20 Mar 2008 (Also: Proceedings of SPIE, vol. 6981. In: Suresh R (ed.) Defense Transformation and Net-Centric Systems 2008, 69810T, 3 Apr 2008)
34. Sapaty, P.S.: Distributed technology for global dominance. Keynote lecture. In: Proceedings of the Fifth International Conference in Control, Automation and Robotics ICINCO 2008, The Conference Proceedings, Funchal, Madeira, Portugal, 11–15 May 2008
35. Sapaty, P.S.: Human-robotic teaming: a compromised solution. In: AUVSI's Unmanned Systems North America 2008, San Diego, USA, 10–12 June 2008
36. Sapaty, P.S.: High-level communication protocol for dynamically networked battlefields. In: Proceedings of International Conference Tactical Communications 2009 (Situational Awareness and Operational Effectiveness in the Last Tactical Mile), One Whitehall Place, Whitehall Suite and Reception, London, UK (2009)
37. Sapaty, P.S.: Distributed capability for battlespace dominance. In: Electronic Warfare 2009 Conference and Exhibition. Novotel London West Hotel and Conference Center, London, 14–15 May 2009
38. Sapaty, P.S.: Providing spatial integrity for distributed unmanned systems. In: Proceedings of 6th International Conference in Control, Automation and Robotics ICINCO 2009, Milan, Italy (2009)
39. Sapaty, P.S.: Distributed technology for global control. Book chapter, Lecture Notes in Electrical Engineering, Informatics in Control, Automation and Robotics, vol. 37. Springer, Berlin (2009)
40. Sapaty, P.S.: Gestalt-based integrity of distributed networked systems. In: SPIE Europe Security + Defence. bcc Berliner Congress Centre, Berlin, Germany (2009)
41. Sapaty, P.S.: Remote control of open groups of remote sensors. In: Proceedings of SPIE Europe Security + Defence, Berlin, Germany (2009)

42. Sapaty, P.S.: Tactical communications in advanced systems for asymmetric operations. In: Proceedings of Tactical Communications 2010, CCT Venues, Canary Wharf, London, UK, 28–30 Apr 2010
43. Sapaty, P.S.: High-level technology to manage distributed robotized systems. In: Proceedings of Military Robotics 2010, London, UK, 25–27 May 2010
44. Sapaty, P.S.: Emerging asymmetric threats, Q&A session. In: Tactical Communications 2010, CCT Venues, Canary Wharf, London, UK, 28–30 Apr 2010
45. Sapaty, P.S.: High-level organization and management of directed energy systems. In: Proceedings of Directed Energy Weapons 2010, CCT, Canary Wharf, London, UK, 25–26 Mar 2010
46. Sapaty, P.S.: Unified transition to robotized armies with spatial grasp technology. In: International Summit Military Robotics, London, United Kingdom, 12–13 Nov 2012
47. Sapaty, P.S.: Distributed air and missile defense with spatial grasp technology. Intell. Control Autom. Sci. Res. **3**(2), 117–131 (2012)
48. Sapaty, P.S.: Global electronic dominance. In: 12th International Fighter Symposium, 6th–8th Nov 2012, Grand Connaught Rooms, London, UK
49. Sapaty, P.S.: Providing global awareness in distributed dynamic environments. In: International Summit ISR, London, 16–18 Apr 2013
50. Sapaty, P.S.: Ruling distributed dynamic worlds with spatial grasp technology. Tutorial at the International Science and Information Conference 2013 (SAI), London, UK, 7–9 Oct 2013
51. Sapaty, P.S.: Night vision under advanced spatial intelligence: a key to battlefield dominance. In: International Summit Night Vision 2013, London, 4–6 June 2013
52. Sapaty, P.S.: Integration of ISR with advanced command and control for critical mission applications. In: SMi's ISR Conference, Holiday Inn Regents Park, London, 7–8 Apr 2014
53. Sapaty, P.S.: Unified transition to cooperative unmanned systems under spatial grasp paradigm. Int. J. Trans. Netw. Commun. **2**(2) (2014)
54. Sapaty, P.S.: From manned to smart unmanned systems: a unified transition. In: SMi's Military Robotics, Holiday Inn Regents Park, London, 21–22 May 2014
55. Sapaty, P.S.: Unified transition to cooperative unmanned systems under spatial grasp paradigm. In: 19th International Command and Control Research and Technology Symposium, Alexandria, Virginia, 16–19 June 2014
56. Sapaty, P.S.: Distributed human terrain operations for solving national and international problems. Int. Relat. Dipl. **2**(9) (2014)
57. Sapaty, P.S.: Providing over-operability of advanced ISR systems by a high-level networking technology. In: SMI's Airborne ISR, Holiday Inn Kensington Forum, London, United Kingdom, 26–27 Oct 2015
58. Sapaty, P.S.: Distributed missile defence with spatial grasp technology. In: SMi's Military Space, Holiday Inn Regents Park, London, 4–5 Mar 2015
59. Sapaty, P.S.: Military robotics: latest trends and spatial grasp solutions. Int. J. Adv. Res. Artif. Intell. **4**(4) (2015)
60. Sapaty, P.S.: A brief introduction to the spatial grasp language (SGL). J. Comput. Sci. Syst. Biol. **09**(02) (2015). https://www.researchgate.net/publication/307744279_A_Brief_Introduction_to_the_Spatial_Grasp_Language_SGL
61. Sapaty, P.S.: Organization of advanced ISR systems by high-level networking technology. Math. Mach. Syst. (1) (2016)
62. Sapaty, P.S.: Towards massively robotized systems under spatial grasp technology. J. Comput. Sci. Syst. Biol. **9**(1) (2016)
63. Sapaty, P.S.: Towards wholeness and integrity of distributed dynamic systems. J. Comput. Sci. Syst. Biol. **9**(3) (2016)
64. Sapaty, P.S.: Towards global goal orientation, robustness and integrity of distributed dynamic systems. J. Int. Relat. Dipl. **4**(6) (2016)
65. Sapaty, P.: Spatial grasp language (SGL). In: Advances in Image and Video Processing, vol. 4, no. 1 (2016). https://journals.scholarpublishing.org/index.php/AIVP/article/view/1922/1052

66. Sapaty, P.S.: Mosaic warfare: from philosohpy to model to solution. Mathematical Mach. Syst. (3) (2019)
67. Sapaty, P.S.: Global network management under spatial grasp paradigm. Int. Robot. Autom. J. **6**(3), 134–148 (2020). https://medcraveonline.com/IRATJ/IRATJ-06-00212.pdf
68. Sapaty, P.S. :Global network management under spatial grasp paradigm. Glob. J. Res. Eng. J Gen. Eng. **20**(5), 58–81. Version 1.0 (2020). https://globaljournals.org/GJRE_Volume20/6-Global-Network-Management.pdf
69. Sapaty, P.: Advanced terrestrial and celestial missions under spatial grasp technology. Aeronaut. Aerosp. Open Access J. **4**(3) (2020). https://medcraveonline.com/AAOAJ/AAOAJ-04-00110.pdf
70. Sapaty, P.: Spatial management of distributed social systems. J. Comput. Sci. Res. **02**(03) (2020). https://ojs.bilpublishing.com/index.php/jcsr/article/view/2077/pdf
71. Sapaty, P.: Towards global nanosystems under high-level networking technology. Acta Sci. Comput. Sci. **2**(8) (2020). https://www.actascientific.com/ASCS/pdf/ASCS-02-0051.pdf
72. Sapaty, P.: Symbiosis of distributed simulation and control under spatial grasp technology. SSRG Int. J. Mob. Comput. Appl. (IJMCA) **7**(2) (2020). http://www.internationaljournalssrg.org/IJMCA/2020/Volume7-Issue2/IJMCA-V7I2P101.pdf
73. Sapaty, P.S.: Symbiosis of real and simulated worlds under global awareness and consciousness. In: Abstract, The Science of Consciousness (TSC) Conferences, Tucson, Arizona (2020). https://eagle.sbs.arizona.edu/sc/report_poster_detail.php?abs=3696
74. Sapaty, P.S.: Spatial grasp as a model for space-based control and management systems. Math. Mach. Syst. (1), 135–138 (2021). http://www.immsp.kiev.ua/publications/articles/2021/2021_1/Sapaty_book_1_2021.pdf
75. Sapaty, P.S.: Managing multiple satellite architectures by spatial grasp technology. Math. Mach. Syst. (1), 3–16. http://www.immsp.kiev.ua/publications/eng/2021_1/
76. Sapaty, P.S.: Spatial management of large constellations of small satellites. Math. Mach. Syst. (2) (2021). http://www.immsp.kiev.ua/publications/articles/2021/2021_2/02_21_Sapaty.pdf
77. Sapaty, P.S.: Global management of space debris removal under spatial grasp technology. Acta Sci. Comput. Sci. **3**(7) (2021). https://www.actascientific.com/ASCS/pdf/ASCS-03-0135.pdf
78. Sapaty, P.S.: Space debris removal under spatial grasp technology. Netw. Commun. Technol. **6**(1) (2021). https://www.ccsenet.org/journal/index.php/nct/article/view/0/45486
79. Sapaty, P.S.: Relation of spatial grasp paradigm to higher psychological and mental concepts. Acta Sci. Comput. Sci. **4**(12) (2022). https://actascientific.com/ASCS/pdf/ASCS-04-0359.pdf
80. Sapaty, P.S.: Seeing and managing distributed worlds with spatial grasp paradigm. Acta Sci. Comput. Sci. **4**(12) (2022). https://actascientific.com/ASCS/pdf/ASCS-04-0365.pdf
81. Sapaty, P.S.: Comprehending distributed worlds with the spatial grasp paradigm. Math. Mach. Syst. (1) (2022). http://www.immsp.kiev.ua/publications/articles/2022/2022_1/01_22_Sapaty.pdf
82. Sapaty, P.S.: Providing global awareness in distributed dynamic systems. Int. Relat. Dipl. **11**(2), 87–100 (2023). https://doi.org/10.17265/2328-2134/2023.02.002. https://www.davidpublisher.com/Public/uploads/Contribute/6486c3d05a6cc.pdf
83. Sapaty, P.S.: Simulating distributed consciousness with spatial grasp model. Math. Mach. Syst. (3) (2023)
84. Sapaty, P.S.: Network centrality operations under spatial grasp technology. J. Adv. Artif. Intell. Mach. Learn. **1**(1), 1–11 (2023). https://www.scivisionpub.com/journals/articleinpress-journal-of-advances-in-artificial-intelligence-and-machine-learning
85. Sapaty, P.S.: Managing distributed systems with spatial grasp patterns. Math. Mach. Syst. (4) (2023). http://www.immsp.kiev.ua/publications/articles/2023/2023_4/04_23_Sapaty.pdf
86. Sapaty, P.S.: Spatial management of air and missile defence operations. Math. Mach. Syst. (1) (2023). http://www.immsp.kiev.ua/publications/articles/2023/2023_1/01_23_Sapaty.pdf
87. Sapaty, P.S.: Providing distributed system integrity under spatial grasp technology. Math. Mach. Syst. (2) (2023). http://www.immsp.kiev.ua/publications/articles/2023/2023_2/02_23_Sapaty.pdf

88. Sapaty, P.S.: Spatial networks as models for organoid cultures and brain research. Math. Mach. Syst. (1) (2024). ISSN 1028-9763. http://www.immsp.kiev.ua/publications/articles/2024/2024_1/01_24_Sapaty.pdf
89. Sapaty, P.S.: Global and international security under spatial grasp paradigm. Int. Relat. Dipl. **12**(2), 72–85 (2024). https://doi.org/10.17265/2328-2134/2024.02.004. https://www.davidpublisher.com/index.php/Home/Article/index?id=50834.html
90. Sapaty, P.S.: Networking solutions in combined distributed worlds. In: Proceedings of the 6th International Scientific Conference «European Research Materials», pp. 256–275, 30–31 May 2024, Amsterdam, Netherlands (2024). ISBN 978-1-6484-8208-3. https://doi.org/10.5281/zenodo.11432625. https://ojs.publisher.agency/index.php/ERM/issue/view/85/210
91. Sapaty, P.S.: Self-recovering infrastructures and networks under spatial grasp paradigm. In: Proceedings of the 7th International Scientific Conference «Theoretical Hypotheses and Empirical results», pp. 96–112, 11–12 July 2024, Oslo, Norway (2024). ISBN 978-0-2324-6742-0. https://doi.org/10.5281/zenodo.12741275. https://ojs.publisher.agency/index.php/THIR/issue/view/91/222
92. Sapaty, P.S.: Self-recovering systems under spatial grasp technology. Cris. Stress. Hum. Resil. Int. J. (2024) (in print). E-ISSN:2836-1709. https://www.crisisjournal.org/about

Chapter 5
Representing Physical and Virtual Spaces in SGL

Abstract The Chapter reviews different ideas, features, and publications related to physical, virtual and hybrid spaces, as well as their unity and integration. Representation of different features of distributed spaces is analysed in SGL, and concrete methods and mechanisms of spatial vision, penetration, navigation and management are offered in SGL. These include declaration and management of spatial communities, operations with large distributed networks and their borders, finding perimeters of complex spatial regions by collective navigation of their spaces, hierarchical network organization, as well as possible spatial art. The provided investigations and resulted solutions confirm clarity, efficiency, and compactness of expressing them in SGL.

5.1 Introduction

Physical space is our familiar, tangible world, while ***virtual space*** is an artificial, digitally created environment. These spaces can interact and influence each other, creating ***hybrid spaces*** where digital and physical experiences blend. Virtual space, though not tied to physical location, can affect activity in the physical world and vice versa. These spaces can create a complex interplay between the real and the digital. ***Distributed space*** is a concept where multiple components are spread out in space, working together to achieve a common goal, rather than relying on a single, large component. This approach offers potential advantages in cost, complexity, and mission capabilities, and is enabled by advancements in technologies like in-space networking and autonomous planning.

The aim of this Chapter is to investigate, analyze, understand, and then express in SGL representation of large distributed spaces of physical and virtual nature, also their combination and integration.

The rest of the Chapter reviews different ideas related to physical, virtual and hybrid spaces; offers examples of spatial operations in SGL; shows how to represent and manage communities in SGL; works with large distributed networks and their borders in SGL; finds perimeters of complex spatial regions in SGL; explains

P. S. Sapaty, *Distributed Management with Spatial Grasp Model*, Studies in Systems, Decision and Control 666, https://doi.org/10.1007/978-3-032-19378-0_5

hierarchical network organization in SGL; shows possible spatial art in SGL; and concludes the chapter with citing related publications.

5.2 Using Physical, Virtual and Hybrid Spaces

The following investigates different spatial ideas with references to related publications.

High-dimensional distributed semantic spaces have proven useful and effective for aggregating and processing visual, auditory, and lexical information for many tasks related to human-generated data. Human language makes use of a large and varying number of features of various types, which all interact to represent various aspects of communicative information [1].

Understanding distributed representations is a fundamental concept in the field of machine learning and natural language processing (NLP). They refer to a way of representing data, typically words or phrases, as continuous vectors in a high-dimensional space. Unlike local representations, distributed representations capture a notion of similarity and semantic meaning by allowing an entity to be represented by a pattern of values across many dimensions [2].

The ***mathematical representation of semantics*** is a key issue for Natural Language Processing (NLP). A lot of research has been devoted to finding ways of representing the semantics of individual words in vector spaces. Distributional approaches, meaning distributed representations that exploit co-occurrence statistics of large corpora, have proved popular and successful across a number of tasks [3].

Space representation refers to the encoding of information in k-space, where data is organized in a matrix that reflects spatial frequencies rather than direct spatial positions. Each point in k-space corresponds to the spatial frequency and signal phase of pixels in the final image, facilitating the conversion of raw data into a visual representation through mathematical transformation [4].

Linear Relational Embedding is a method of learning a distributed representation of concepts from data consisting of binary relations between concepts. A representation for concepts and relations is learned by maximizing an appropriate discriminative goodness function using gradient ascent. On a task involving family relationships, learning is fast and leads to good generalization [5].

The ***strongest convention in contemporary geographic thought*** is the notion that geographic space is rooted in a Euclidean geometry that defines the physical world. Although geographers have long sought to escape this paradigm through a rich array of perceptions based on ways in which we might imagine space, physical distance or its economic surrogates still provide the basic logic used by geographers [6].

Today, the ***opportunity for potential human activity*** has gone beyond physical space to virtual space. Based on a proposed conceptual framework that models the relationships between physical and virtual spaces, this presents an attempt to adjust the space–time concept to identify potential activity opportunities in virtual space, focusing on the virtual space access channels available in physical space [7].

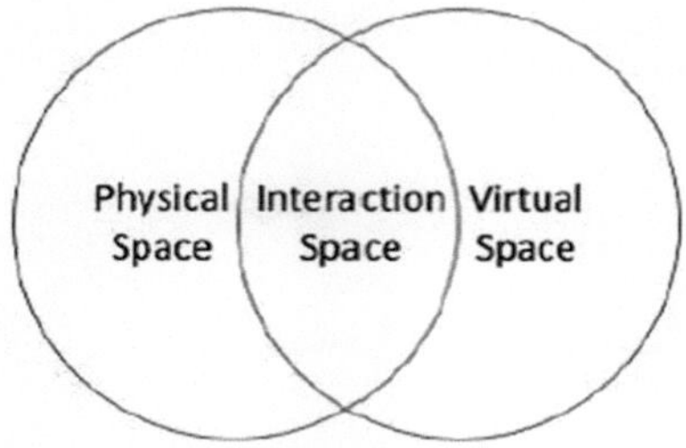

Fig. 5.1 Interaction of physical and virtual spaces

In the ***experience and design of spaces today*** we meet both reality and virtuality. But how is the relation between real and virtual construed? How can we as researchers and designers contribute to resolving the physical-virtual divide regarding spaces? The work explores relations between the physical and the virtual and investigates ways of connecting physical and virtual space, both in theory and practice [8].

After this short review we may generally conclude that physical space refers to the tangible, real-world environment, while virtual space is a computer-generated representation of that space or other environments, which their symbolical interaction in Fig. 5.1.

5.3 Elementary Spatial Operations

This section and the rest of the Chapter will be using different methods and mechanisms expressed in SGL, which may also relate to previous publications on SGL and SGT, like [9–23], also older ones cited in other chapters.

- **For physical world** (Fig. 5.2)

(a) *Isolated node creation* (with name 1 and physical coordinates X1_Y1, as in Fig. 5.2a):

```
create_node(1, X1_Y1)
```

(b) *Extending existing node to elementary network* (with new node 2 having coordinates X2_Y2 and physical link or road to it from the previous node, as in Fig. 5.2b):

```
move_node(1); create(Road, node(2, X2_Y2))
```

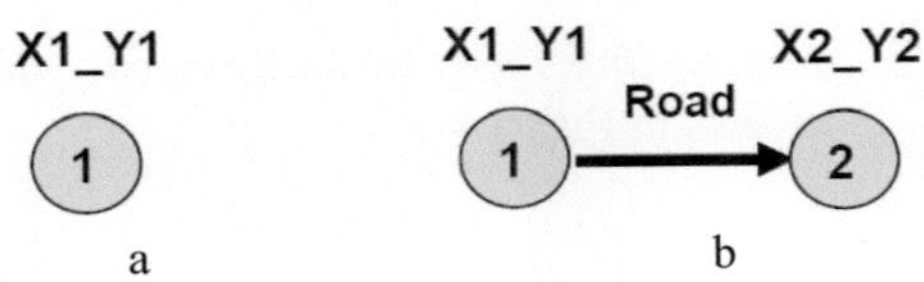

Fig. 5.2 Elementary physical world creation and extension

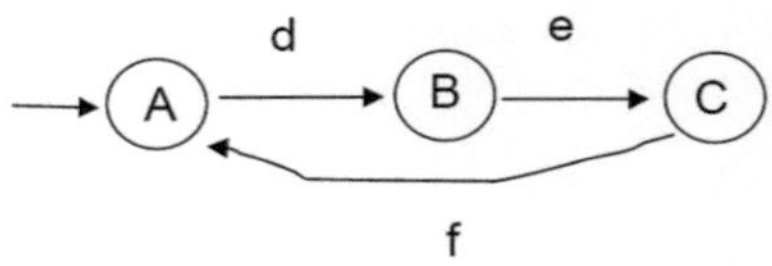

Fig. 5.3 Elementary virtual world creation

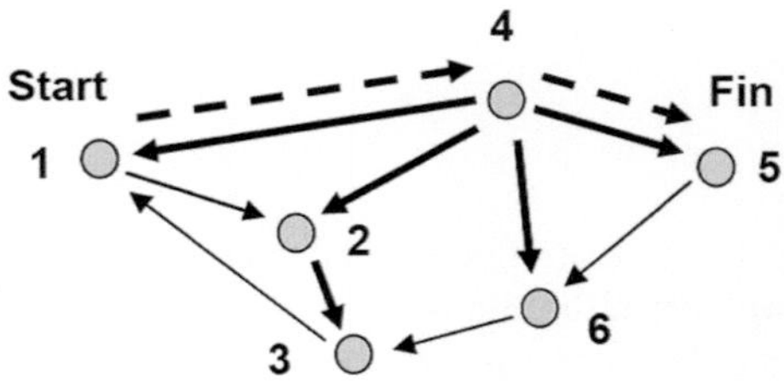

Fig. 5.4 Physical network example

- **For virtual world** (Fig. 5.3)

Full network creation:

```
create(node(A));   create(link(d),   node(B));   create(link(e),
node(C));
linkup(f, node(A))
```

More compact, if needed:

```
create (A; d*B; e*C); linkup(f*A) or
create(A; d*B; e*C; f*!A)
```

where `!A` informs that node with name `A` already exists, and only link to it should be created by `create`.

- **Reaching certain point in a network by shortest path to it (**dashed to Fin from Start, Fig. 5.4)

(a) *Distributed creation of this network* (hard arrows in Fig. 5.4 reflecting initial network growth, thin ones subsequent stages)

```
create_node(4, X4_Y4);
sequence(
   create(link, node(1, X1_Y1)),
  (create(link, node(2, X2_Y2)); create(link, node(3, X3_Y3))),
   create(link, node(6, X6_Y6)),
   (create(link, node(5, X5_Y5)); linkup_node(6);
     Linkup_node(3); linkup_node(1); linkup_node(2)))
```

(b) *First finding, then physically moving via shortest route from Start to Fin* (dashed arrows in Fig. 5.4).

```
nodal(Now = infinite, Route, Start = 1);
frontal(Path, Length, Fin = 5);
move(Start); Path = NAME;
sequence(
   repeat(
   hop(all_links); not_belong(NAME, Path); append(Path, NAME);
      Length += distance(BEFORE, WHERE);
      if(and(NAME == Fin, Length < Now),
         stop(Now = Length; Route = Path))),
   (Path = (hop(Fin); Route);
    repeat(move(withdraw(Path[1])))))
```

- **Converting a curved way to the sequence of straight paths** (Fig. 5.5)

(a) *Stepwise finding new points in the curved road by space search with certain Radius*

```
frontal(Radius = …, Link = …); Start = X_Y;
create_node(Start);
repeat(create(+ Link, node(new_intersect(Radius))));
```

(b) *Movement from Start to Fin via created paths*

```
frontal(Link) = …; Start = X_Y;
move(Start); repeat(move(+ Link));
```

(c) *Direct move from Start to Fin* (if possible by the territory)

```
frontal(Link) = …; Start = X_Y;
move(Start); Fin = (repeat(hop(+ Link)); WHERE);
move(Fin)
```

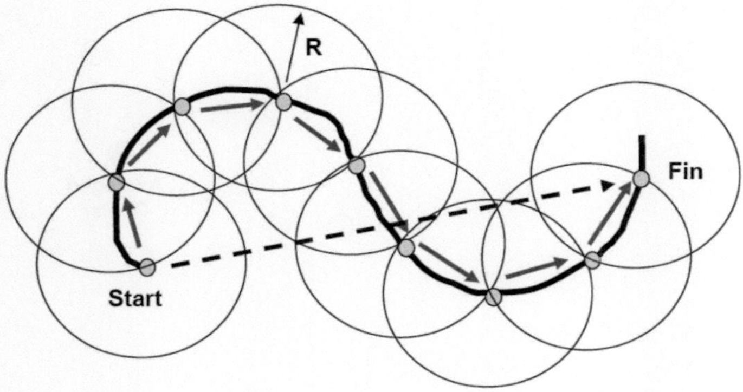

Fig. 5.5 From curved to straight paths

5.4 Representing Communities

- **Finding centre of the community** (Fig. 5.6)

```
frontal(Aver) = average(hop(all_nodes); WHERE);
Center = (fringe_min(hop(all_nodes); distance(WHERE, Aver));
NAME)
```

- **Spanning tree of the community** (Fig. 5.7)

(a) *Creating spanning tree based on certain distances between nodes*

```
Center = …; R = …; hop_first(Center);
repeat(+linkup(hop_first(within(R))))
```

The found centre will be 8.

(b) *Finding path from the centre to a certain node* (like 34)

```
frontal(Path); hop(34);
repeat(Path = append(NAME, Path); hop(-links));
output(Path)
```

Result issued in the central node 8 will be: `8, 20, 35, 4, 34`

- **Creating fully centralized organization** (from the centre to all other nodes, as in Fig. 5.8)

```
Center = …; hop(Center); linkup(hop(all_other_nodes))
```

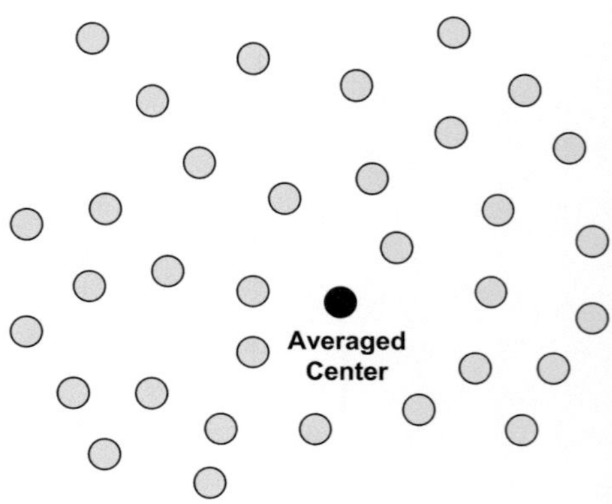

Fig. 5.6 Centre of community

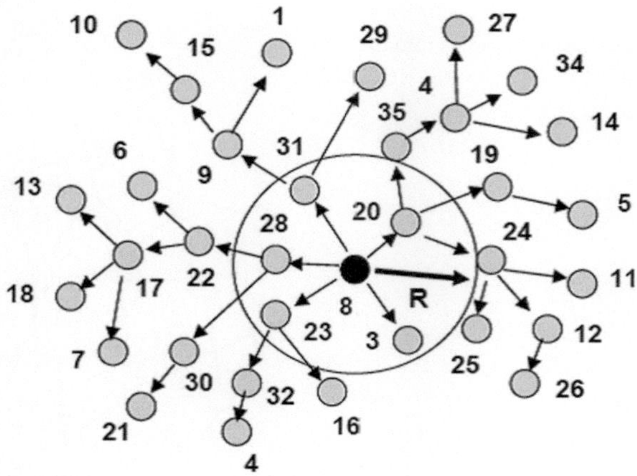

Fig. 5.7 Spanning tree of community

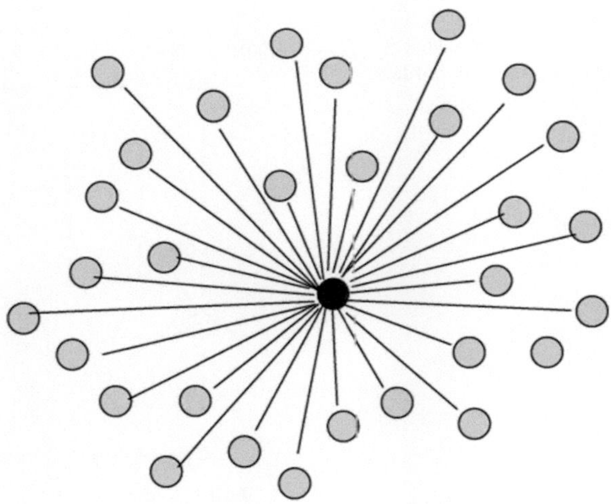

Fig. 5.8 Centralized organization

5.5 Working with Networks

- **Directly linking existing nodes within certain distance R from each other** (Fig. 5.9)

```
R = …; hop_nodes(all); frontal(Prev) = NAME;
hop(within(R)); Prev > NAME; linkup(Prev)
```

- **Finding the network dimensions and surrounding borders creation** (Fig. 5.10)

The resultant borders can also be served altogether as a security fence, if the Border is supplied with proper features.

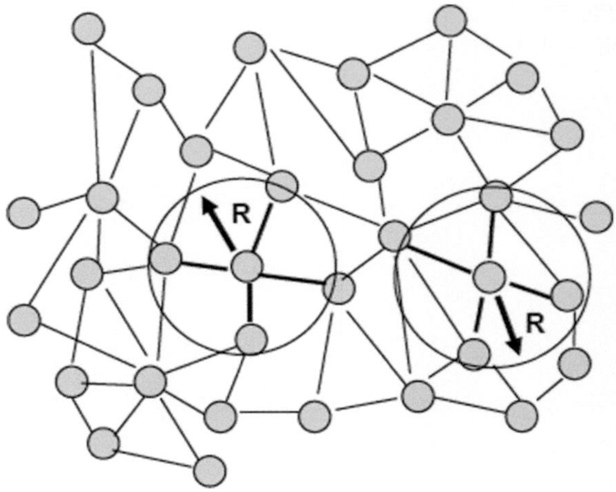

Fig. 5.9 Linking neighbouring nodes into the network

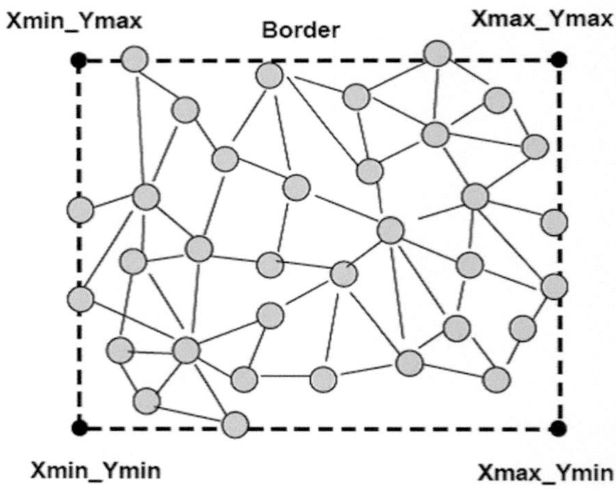

Fig. 5.10 Establishing network security borders

```
frontal(
  Xmin = min(hop(all_nodes); dimension(X, WHERE)),
  Xmax = max(hop(all_nodes); dimension(X, WHERE)),
  Ymin = min(hop(all_nodes); dimension(Y, WHERE)),
  Ymax = max(hop(all_nodes); dimension(Y, WHERE)));
Border = type; move(Xmin_Ymin);
move_linkup(Border, Xmax_Ymin); move_linkup(Border, Xmax_Ymax);
move_linkup(Border, Xmin_Ymax); move_linkup(Border, Xmin_Ymin)
```

- **Finding the network area centre and creating circular border** (Fig. 5.11)

```
frontal(Aver) = average(hop(all_nodes); WHERE);
Center = (fringe_min(hop(all_nodes); distance(WHERE, Aver));
NAME);
```

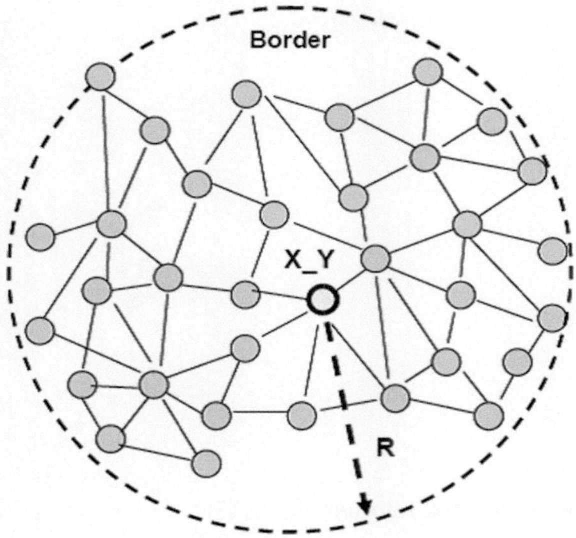

Fig. 5.11 Creating circular border

```
hop(Center); frontal(Coord = WHERE, Border = type);
R = max(hop(other_nodes); distance(Coord, WHERE));
create_circular_border(Border, radius(R))
```

5.6 Finding Perimeters of Complex Regions

Finding perimeter coordinates of complex spatial regions can be effectively organized in SGL in different ways, some examples following.

- **Parallel solution**

The following parallel scenario after initial landing inside the region of interest, replicates, spreads and replicates again until reaches regional borders, with collecting coordinates of first points which appeared just outside the region's space, as in Fig. 5.12.

```
hop(any, TYPE == in);
frontal(Copies = number, Depth = dist);
Border = repeat(max)(replicate(current, Copies);
             hop(angle_random(360), Depth);
             if(TYPE == out, done(WHERE)))
```

Result: `X1_Y1, X2_Y2, X3_Y3, …, Xn_Yn`

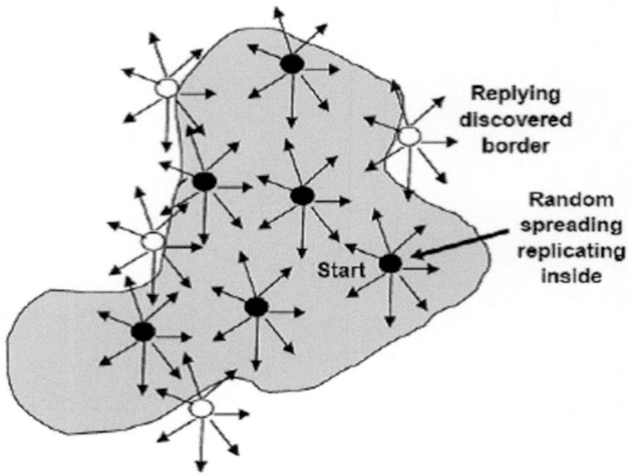

Fig. 5.12 Parallel solution for the perimeter

- **Sequential solution** (Fig. 5.13)

The following scenario, after initial landing just nearby the region's perimeter, makes a sequence of close outside-inside hops, believably enough to trace the whole perimeter, while accumulating coordinates of all these hops, as in Fig. 5.13.

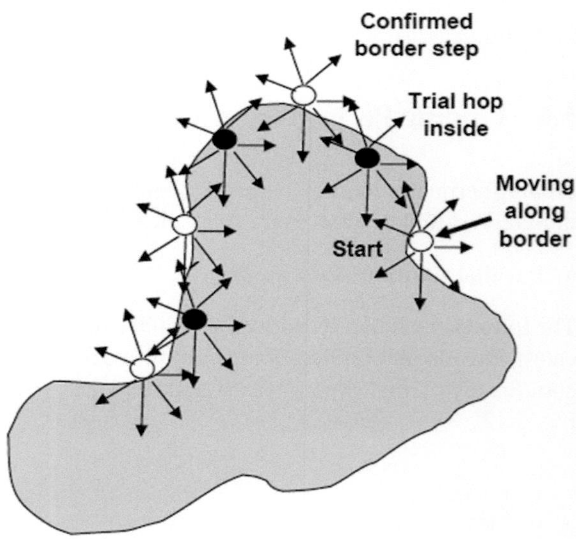

Fig. 5.13 Sequential solution for the perimeter

```
hop(border_close), TYPE == out);
frontal(Border = WHERE, Depth = dist);
repeat(max)(
   loop_unless(hop(angle_random(360), Depth); TYPE == in);
   append(Border, WHERE);
    loop_unless(hop(angle_random(360), Depth); TYPE == out);
   append(Border, WHERE));
output(Border)
```

Result: `X1_Y1, X2_Y2, X3_Y3, …, Xn_Yn`

- **Working with more complex areas** (Fig. 5.14)

For more complex regions, like in Fig. 5.14, a combination of their navigation in parallel and sequential mode may be needed, with initial massive landing inside the marked region area (outlined by certain Dimensions), and then, when reaching the region borders, move only along them with collecting the closest to border coordinates. Only such strategy can guarantee all borders to be discovered and accumulated, as follows.

```
hop(massive, Dimensions = (X1, X2, Y1, Y2), TYPE == in);
frontal(Copies = number, Depth = dist);
repeat(max1)
   (replicate(current, Copies);
   hop(angle_random(360), Depth);
   if(TYPE == out,
      (frontal(Border = WHERE, Depth = dist);
       repeat(max2)(
          loop_unless(hop(angle_random(360), Depth); TYPE == in);
          append(Border, WHERE);
         loop_unless(hop(angle_random(360), Depth); TYPE == out);
          append(Border, WHERE));
       output(Border)))
```

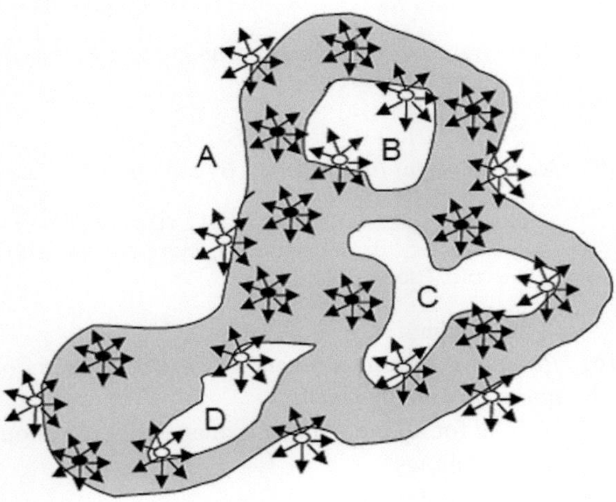

Fig. 5.14 Parallel-sequential perimeter solution for complex areas

Fig. 5.15 Solutions as sets of dots: **a** for Figs. 5.12 and 5.13, and **b** for Fig. 5.14

Result:

Perimeter A: XA1_YA1, XA2_YA2, XA3_YA3, …, XAn_YAn
Perimeter B: XB1_YB1, XB2_YB2, XB3_YB3, …, XBn_YBn
Perimeter C: XC1_YC1, XC2_YC2, XC3_YC3, …, XCn_YCn
Perimeter D: XD1_YD1, XD2_YD2, XD3_YD3, …, XDn_YDn

The above solutions for the perimeters of complex areas were obtained as the sets of dots on them with proper coordinates (symbolically in Fig. 5.15), with the results more detailed if higher values for *number*, *massive*, *max*, *max1*, and *max2*, also lower for *dist* were used.

5.7 Hierarchical Network Organizations

An example of hierarchical network is shown in Fig. 5.16 consisting of interlinked management and production networks. Some elementary queries to this organization may be as follows.

(a) *Finding total number of production nodes-units*

```
Number = count(hop_nodes(m1, m2, m3, m4); hop_links(subord));
output(Number)
```

Result: `17`

(b) *Most powerful organization by the number of production nodes*

```
Most = max(hop_nodes(m1, m2, m3, m4);
           append(count(hop_links(subord)), NAME);
output(Most[2]);
```

Result: `m4`

(c) *Providing locations awareness for all production units about all units under another management*

Like for all units under management m3 about all units under management m1, as follows.

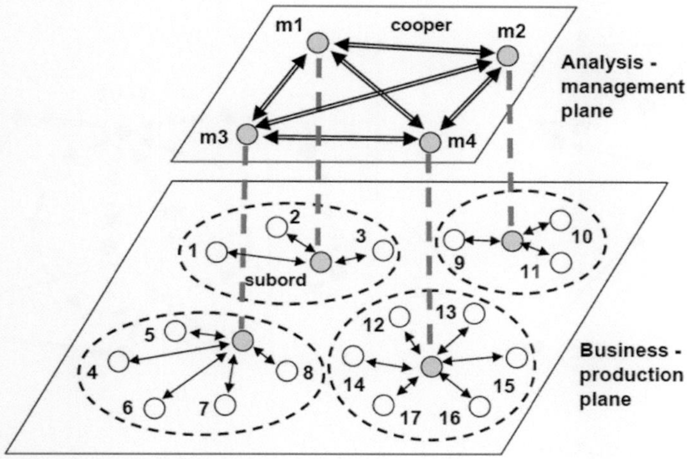

Fig. 5.16 Hierarchical networking

```
hop_node(m1); frontal(List);
List = append(NAME, unit(hop_links(subord); NAME & WHERE));
hop(link(cooper), node(m3)); hop_links(subord); Collabor =
List
```

Each node subordinate to m3 (i.e.: 4, 5, 6, 7, and 8) will have now the full list of collaborator nodes from m1 in the individual variable Collabor, as follows:

```
m1(1*X1_Y1, 2*X2_Y2, 3*X3_Y3)
```

5.8 Spatial Art

Elementary examples are shown in Fig. 5.17.

(a) *Creating virtual image* (of Fig. 5.17a).

```
create((1, blue); (link, (2, blue)); (link, 3);
linkup(1),
create((link, 4); (link, 5), (link, 6),
       ((link, 7); (link, 8), (link, 9)))
```

(b) *Creating physical image* (of Fig. 5.17b).

```
frontal(Start = X_Y_Z, Type, Dev = dx_dy_dz, Steps = 5);
create(Start);
split(Type = (dotted, black),
      Type = (dashed, blue),
      Type = (double, red));
```

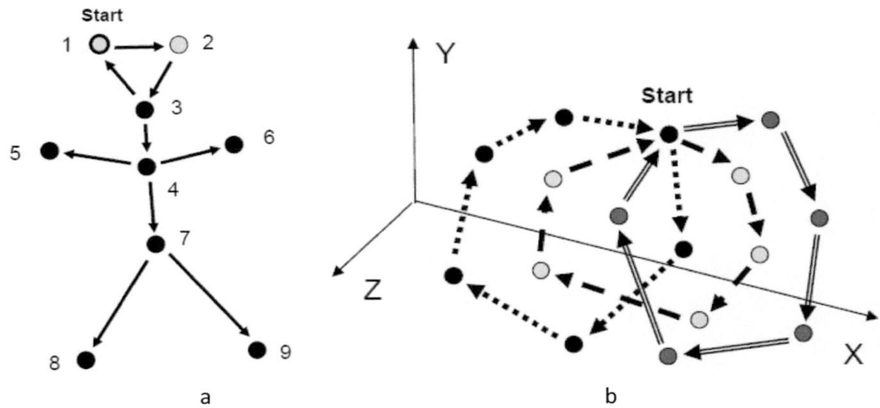

Fig. 5.17 Spatial art examples

```
repeat(Steps)(
  create(link(Type[1]), node(shift_random(Dev), Type[2]))));
linkup(Type[1], Start)
```

5.9 Conclusions

After reviewing different ideas and concepts related to physical, virtual and hybrid spaces, the Chapter provided examples of dealing with various spaces under SGT, including their description, modification, and solving different problems in them, which can confirm clarity, efficiency, and compactness of expressing them in SGL.

References

1. Karlgren, J., Kanerva, P.: High-dimensional distributed semantic spaces for utterances. Computer Science > Computation and Language, Submitted on 1 Apr 2021. https://arxiv.org/abs/2104.00424
2. Understanding distributed representations. https://deepai.org/machine-learning-glossary-and-terms/distributed-representation
3. Hermann, K.M.: Distributed representations for compositional semantics. Computer Science > Computation and Language, Submitted on 12 Nov 2014. https://arxiv.org/abs/1411.3146
4. Space representation. In: Subject Area: Engineering. AI Generated Definition Based on: Encyclopedia of Behavioral Neuroscience, 2nd edn. (2022). https://www.sciencedirect.com/topics/engineering/space-representation
5. Paccanaro, A., Hinton, G.E.: Learning Distributed Representations by Mapping Concepts and Relations into a Linear Space. Gatsby Computational Neuroscience Unit, UCL, London. https://www.cs.utoronto.ca/~hinton/absps/icml-lre.pdf

6. Batty, M., Miller, H.J.: Representing and visualizing physical, virtual and hybrid information spaces. In: Information, Place, and Cyberspace. Advances in Spatial Science. Springer, Berlin, Heidelberg (2000). https://doi.org/10.1007/978-3-662-04027-0_8
7. Yu, H., Shaw, S.-L.: Exploring potential human activities in physical and virtual spaces. Int. J. Geogr. Inf. Sci. **22**(4), 409–430 (2008). https://doi.org/10.1080/13658810701427569
8. Wideström, J.: Connecting Physical and Virtual Spaces. Department of Computer Science and Engineering, Chalmers University of Technology, Gothenburg, Sweden (2020). https://research.chalmers.se/publication/516761/file/516761_Fulltext.pdf
9. Sapaty, P.S.: A distributed processing system. European Patent N 0389655, Publ. 10.11.93. European Patent Office
10. Sapaty, P.S.: Mobile Processing in Distributed and Open Environments. Wiley, New York (1999)
11. Sapaty, P.S.: Ruling Distributed Dynamic Worlds. Wiley, New York (2005)
12. Sapaty, P.S.: Managing Distributed Dynamic Systems with Spatial Grasp Technology. Springer (2017)
13. Sapaty, P.S.: Holistic Analysis and Management of Distributed Social Systems. Springer (2018)
14. Sapaty, P.S.: Complexity in International Security: A Holistic Spatial Approach. Emerald Publishing (2019)
15. Sapaty, P.S.: Symbiosis of Real and Simulated Worlds under Spatial Grasp Technology. Springer (2021)
16. Sapaty, P.S.: Spatial Grasp as a Model for Space-based Control and Management Systems. CRC Press (2022)
17. Sapaty, P.S.: The Spatial Grasp Model: Applications and Investigations of Distributed Dynamic Worlds. Emerald Publishing (2023)
18. Sapaty, P.S.: Providing Integrity, Awareness, and Consciousness in Distributed Dynamic Systems. CRC Press (2024)
19. Sapaty, P.S.: Spatial Networking in the United Physical, Virtual, and Mental World. Springer (2024)
20. Sapaty, P.S.: Self-Healing and Self-Recovering Systems under the Spatial Grasp Model. Emerald Publishing Limited (2025)
21. Sapaty, P.S.: Global and international security under spatial grasp paradigm. Int. Relat. Dipl. **12**(2), 72–85. https://doi.org/10.17265/2328-2134/2024.02.004
22. Sapaty, P.S.: Self-recovering-systems-under-spatial-grasp-technology. Cris. Stress. Hum. Resil. Int. J. **6**(2) (2024). https://www.crisisjournal.org/article/126490-self-recovering-systems-under-spatial-grasp-technology
23. Sapaty, P.S.: Spatial grasp model for distributed management and its comparison with traditional algorithms. Int. Relat. Dipl. **13**(3) (2025). https://doi.org/10.17265/2328-2134/2025.03.004. https://www.davidpublisher.com/Public/uploads/Contribute/689ada551780e.pdf

Chapter 6
Examples of Spatial Management in Different Areas

Abstract The Chapter reviews different spatial management issues by existing publications, and then offers and explains concrete tools and scenarios for spatial management in SGL. These include finding and using shortest path tree and spatial pattern matching in networks, worldwide search for certain individuals identified by specific features and assisted by distributed security systems, chasing and fighting multiple targets at a battlefield under global awareness supported by the chasers regularly informing each other about the targets seen. Other scenarios are for tracking mobile objects by distributed radar networks, cooperative robotic finding of oil spill center, discovering, surrounding, and battling forest fires, also cooperation of autonomous vehicles within optimized lane manoeuvring.

6.1 Introduction

Spatial management encompasses the various methods and approaches used to plan, organize, and regulate the use of physical space, including land, water, and air. It involves a range of activities from strategic planning to the implementation of regulations, all aimed at achieving sustainable and efficient spatial development. This includes balancing competing demands, optimizing resource allocation, and ensuring a high quality of life for present and future generations. ***Spatial operations*** are a class of operations performed on spatial data, which can include vector and raster datasets, to analyze and manipulate their geometric and attribute information. These operations are fundamental to geographic information systems (GIS) and spatial analysis, enabling tasks such as overlay, proximity analysis, and geometric transformations. ***Spatial management operations*** encompass the processes involved in planning, analyzing, and managing the use of physical space, including land, resources, and infrastructure. These operations involve a range of activities, such as spatial planning, spatial data management, and the application of spatial analysis techniques. They are crucial for achieving sustainable development and optimizing resource allocation.

P. S. Sapaty, *Distributed Management with Spatial Grasp Model*, Studies in Systems, Decision and Control 666, https://doi.org/10.1007/978-3-032-19378-0_6

The aim of this Chapter is to review existing ideas and concepts of spatial management in different areas and propose concrete and practical solutions of various management problems with the use of SGT and SGL.

The rest of the Chapter reviews different spatial management issues by existing publications; offers network-based tools for spatial management in SGL like shortest path tree and spatial pattern matching; shows how to trace and find certain individuals worldwide; provides fighting of multiple targets under global awareness supported by regular cooperation of chasers; shows tracking of mobile objects by distributed radar network; describes cooperative robotic finding of oil spill center; investigates surrounding and battling of forest fires; explains optimized collective behaviour of vehicles during their lane manoeuvring; concludes the presentation and cites used references.

6.2 Different Spatial Management Issues

Some ideas with references to related publications, as follows.

Spatial operations use geometry functions to take spatial data as input, analyze the data, then produce output data that is the derivative of the analysis performed on the input data. The analysis performed on the input data returns the coordinates or text representation of the resultant geometries. You can use that information as part of a larger query to perform further analysis, or you can use the results as input to another table [1].

Map operations refer to the operations that an end user performs on maps stored in a database. Map information is stored according to themes (for instance, cities, roads, or population), sometimes called layers in the GIS terminology. The maps considered here are stored in a vector format—as opposed to a raster format such as a grid of pixels [2].

An ***integrated methodology*** has been designed to assist sustainable decision-making processes managed by spatial planners and developers in public authorities and the private sector. It emphasises public participation at early stages of spatial planning for the integration of sustainability criteria. It also acts to validate and complement strategic environmental assessment [3].

Spatial database management deals with the storage, indexing, and querying of data with spatial features, such as location and geometric extent. Many applications require the efficient management of spatial data, including Geographic Information Systems, Computer Aided Design, and Location Based Services. The goal is to provide an overview of spatial data management technology [4].

Preserving natural and semi-natural areas has become a crucial consideration for policymakers, with several drivers recognized as pivotal forces that shape landscapes globally. Among these, socioeconomic, demographic, climatic, and political factors have the most significant implications for landscape changes, contributing to land fragmentation, biodiversity and habitat loss, and overall land degradation [5].

The threats in the twenty-first century demand new innovative tactics, technologies, and doctrine. The time to introduce distributed operations is now. We must continue to develop and implement the concept of distributed operations. The joint force must continue to evolve and leverage the asymmetrical strengths, primarily its technological advances and superbly trained forces [6].

Distributed Maritime Operations is the Navy's service warfighting concept. However, the concept suffers from a variety of interpretations and needs more specificity regarding on what warfighting approaches it is concentrating. While the concept describes mass fires and decision advantage as core themes, it lacks sufficient coherence and concrete focus to effectively guide the Navy's development [7].

The work describes applications of the patented and internationally tested Spatial Grasp Technology (SGT) and its Spatial Grasp Language (SGL) for ***integrated air and missile defence*** (IAMD). Based on holistic space navigation and processing by recursive mobile code self-spreading in distributed words, SGT differs radically from the traditional management of large systems [8].

The following concrete examples of spatial management are using Spatial Grasp Technology and its Spatial Grasp Language described in [8–23] (older sources are cited in other chapters), also explained in detail in Chaps. 3 and 4.

6.3 Network-Based Operations for Spatial Management

The following network-based solutions of finding shortest path tree and matching distributed network with complex patterns may prove important for distributed management systems. The first can be used for optimal top-down and bottom-up management hierarchies while the second for different collective production, management, and security solutions.

- **Creating and using shortest path tree (SPT)**

Imagine we have a distributed physical or virtual network, as in Fig. 6.1.

To create on this network SPT structure starting from some node d and providing shorted paths from it to all other nodes (shown in Fig. 6.1 with hard arrows) we may write in SGL the following scenario (the found SPT will be embedded in the network structure with variables Up in all nodes)

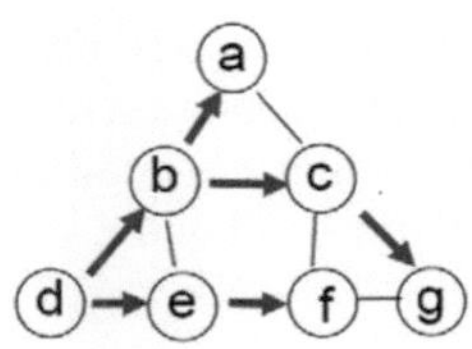

Fig. 6.1 A network with shortest path tree on it

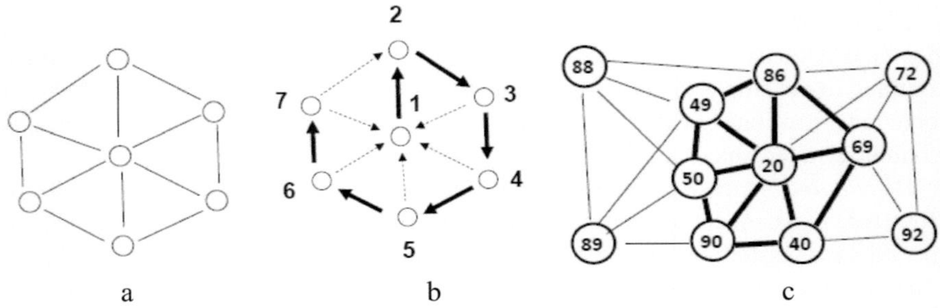

Fig. 6.2 **a** Graph image; **b** its active template for finding exact match; **c** the match found in the network

```
nodal(Dist, Up); hop_node(d); Distance = 0; frontal(Far);
repeat(
  hop_links(all); Far += 1;
  or(Distance == nil, Distance > Far);
  Distance = Far; Up = BACK)
```

After its completion, we may start in any node, say c, and follow contents of nodal variables Up leading to the above nodes of the obtained tree, finally receiving the shortest path to this node from the starting node d as follows:

```
hop_node(c); frontal(Spath) = NAME;
repeat(hop(Up); append(NAME, Spath));
output(Spath)
```

The output result will be `(d,b,c)`.

- **Using spatial pattern matching**

The following SGL scenario is finding exact match of the given image in a distributed network, as in Fig. 6.2.

```
hop_first_nodes(all); COLOR = Match = NAME;
hop_first_links(all); Match & NAME;
repeat(5)(hop_first_links(all); Match & NAME;
     true(hop_links(any, node(Match[1])));
hop_links(any, node(match[2]); output(Match)
```

One of the output results will be: 20, 86, 69, 40, 90, 50, and 49. Having received this set of nodes, say, oriented to represent certain community, we may also additionally create its internal management structure by the scenario similar to the previous SPT example, thus combining spatial pattern matching with hierarchical management and control within the single solution.

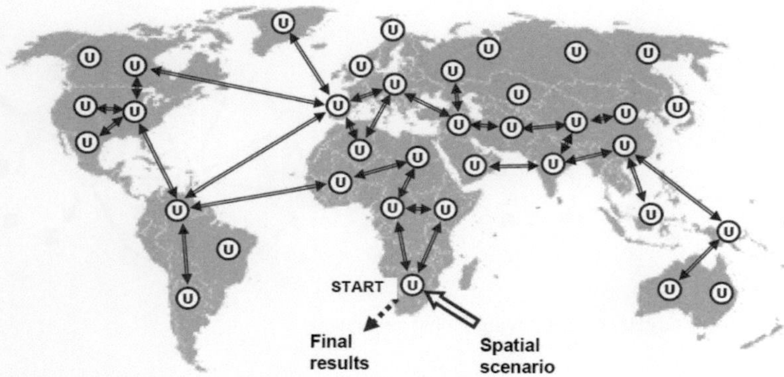

Fig. 6.3 Spatial worldwide search and collection for certain individuals

6.4 Finding Individuals Worldwide

Imagine we want to find detailed information about individuals belonging to some Group indentified by specific features while originating in some START position represented by physical or virtual address. Staying in it, the group members can be found by their features in local databases. This may fail for some or all individuals sought, but their traces may exist in local security systems. If such traces exist and lead to other world locations, we may search both data and security records at these points too, and so on, potentially covering the whole world. The found match from different points with whereabouts of individuals can be returned to START and output there. All this can be expressed in SGL as follows, also symbolically in Fig. 6.3 (with universal SGT interpreters U supposedly installed in many points).

```
hopfirst(START); nodal(Other);
frontal(Group) = features;
output('Records found worldwide:' &&
     repeat(free(match(Group, local_databases)),
          Other = traces(Group, local_security);
          hopfirst(Other)))
```

This scenario first employs spatial forward parallel world propagation, and then in a feedback mode brings the obtained results.

Answer in START may be: *Records found worldwide: (match_1, match_2, ..., match_m)*

6.5 Fighting Targets Under Global Awareness

By regularly enriching the distributed swarm of chasers with a sort of global awareness over the operational area we may essentially improve its performance. This awareness can be effectively embedded into communicating chasers where the targets

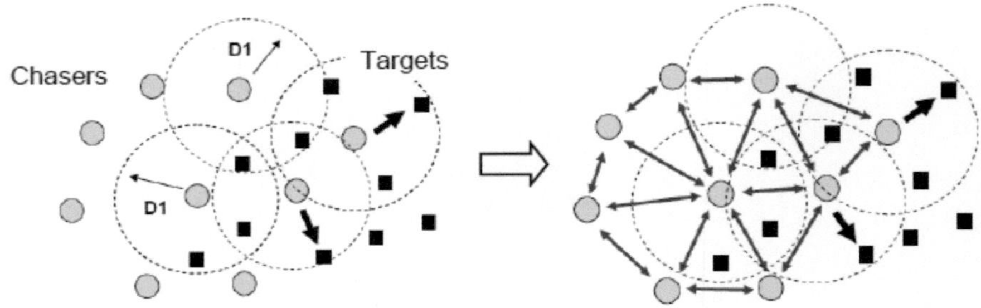

Fig. 6.4 Fighting targets under growing global awareness

seen by individual chasers can be regularly exchanged with their neighbors, as in Fig. 6.4.

This makes all chasers gradually aware of all targets in the region despite not all directly visible individually (D1 as vision threshold), always organizing their movement toward the targets, and then select, move, and destroy them if close enough. An example of SGL solution may be as follows.

```
hop_chasers(all);
nodal(D1 = distance, Targets); frontal(Exchange);
repeat(
   extend(Targets) = search(D1);
   select_move_destroy_remove(Targets);
   stay(Exchange = Targets; hop(neighbors);
      merge(Targets, Exchange));
   sleep(Delay))
```

6.6 Tracking Mobile Objects

The following self-evolving spatial scenario tracks and controls complex object propagating via distributed network of radar stations, as in Fig. 6.5.

```
hop(all_nodes); frontal(Object);
whirl(
   Object = search(aerial, new));
   visibility(Object) > threshold;
   repeat(
      loop(visibility(Object) > Threshold);
      max_destination(hop(all_neighbors); visibility(Object));
      visibility(Object) > Threshold)))
```

This scenario can be easily extended with investigation and detailing of the object's parameters at each step, in addition to their visibility level, which when becomes too dangerous, should be transferred to the distributed network of shooters for the appropriate job.

Fig. 6.5 Distributed objects tracking by a radar network

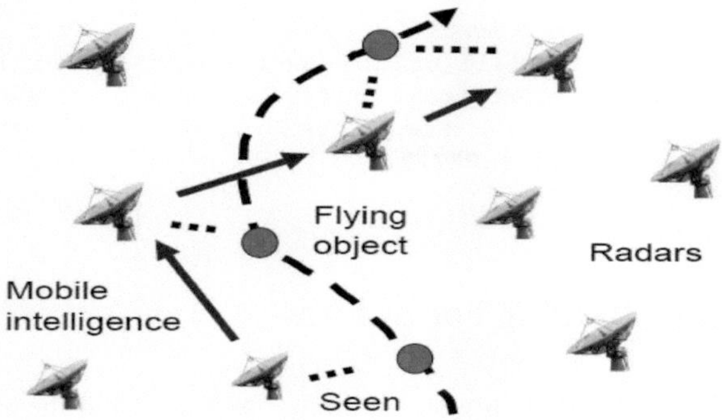

6.7 Cooperative Finding of Oil Spill Center

This scenario describes the situation where a hypothetical region of the sea (surface or submersed) is polluted with spilling oil, and a group of communicating surface or underwater robots, initially delivered to the approximate region, is trying to detect the spill center cooperatively, as in Fig. 6.6. Each unit, if initially found pollution level above certain minimum, tries to move stepwise toward increased spill level in a randomized way (within certain range). After certain time (restricted by allowed max repetitions for each unit), two possible solutions may be considered.

(a) The probable spill center location is defined as an average of final coordinates of all participated units:

```
Center = average(
   hop(all_units); Level = check(spill) > min;
   repeat(max)(or((random(move, range); New = check(spill);
            New > Level; Level = New), stay)); WHERE);
output(Center)
```

(b) The probable spill center is approximated as location of the unit with highest spill registered by it, the latter issuing the result autonomously:

Fig. 6.6 Finding oil spill center

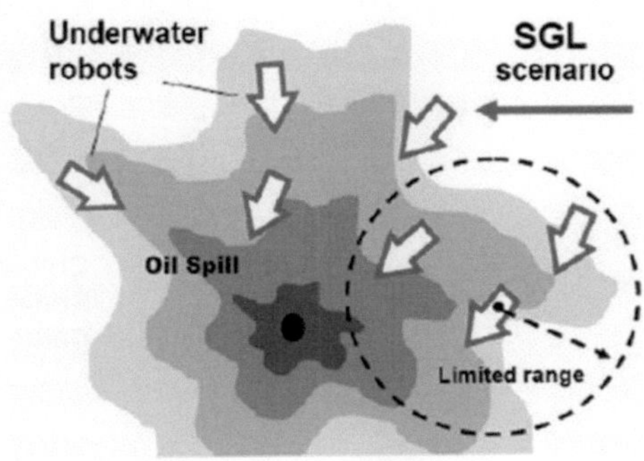

```
fringe_max(
  hop(all_units); Level = check(spill) > min;
  repeat(max)(or((random(move, range); New = check(spill);
        New > Level; Level = New), stay)); Level);
output(WHERE)
```

6.8 Battling Forest Fires

Fighting forest fires is a complex problem, and the use of drones may prove very effective. We will consider a solution where distributed physical space is randomly searched by simultaneous propagation of multiple reconnaissance units. These, after discovering irregularities (like forest fires) encircle the fire zones, collect their perimeter coordinates, transfer them to headquarters (HQ), which initiate massive impact on the zones under fire, as in Fig. 6.7. This impact may be manned, unmanned, or mixed. The SGL scenario for this task may be as follows:

```
move(HQ); create_units(1, 2, 3, 4, 5, 6);
frontal(Zone, Threshold = …);
repeat(max)(
  shift_random(limits);
  if(check( fire),
    (Zone = WHERE; repeat(
      move_around(fire, depth, clockwise); append(Zone, WHERE);
      if(distance(Zone[1], WHERE, anticlockwise) < Threshold,
        (hop_direct(HQ); impact(Zone, global)))))))
```

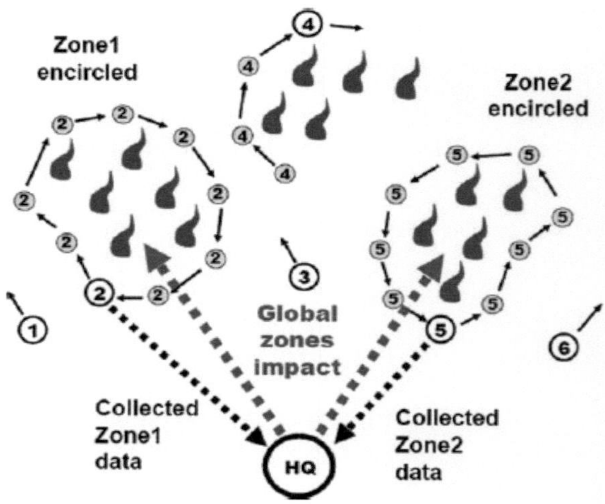

Fig. 6.7 Battling forest fires

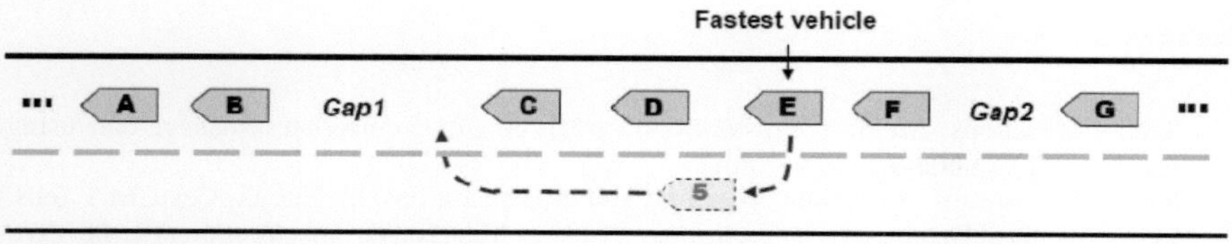

Fig. 6.8 Vehicles collective management

6.9 Autonomous Vehicles Collective Behaviour

We will consider here an exemplary situation on a road like *lane manoeuvring for the fastest vehicle between two gaps*. This situation is shown in Fig. 6.8 and by SGL scenario that follows, where starting from the first vehicle after the gap the whole chain of them is autonomously analyzed and the fastest vehicle found. The latter then making this manoeuvre individually, directly cooperating with the first vehicle it wants to be ahead of, and this can be done manually by a human driver or fully automatically by a robotic vehicle. This manoeuvring scenario will be working in all vehicles endlessly during their continuous cooperative movement in the lane.

```
hop(all_lane_vehicles);
frontal(Gap = …, Max, First, Chosen);
sling(
   sleep(delay);
   distance(ahead) > Gap; First = ADDRESS;
   repeat(
      if(MAXSPEED > Max, (Max = MAXSPEED; Chosen = ADDRESS));
      distance(behind) < Gap; hop(next_behind));
   hop(Chosen); lane_manoeuvre(ahead, First))
```

6.10 Conclusions

After reviewing different spatial management ideas and issues by existing publications, the Chapter offered spatial management solutions expressed by the SGL scenarios which showed clarity, effectiveness, and compactness of the use of SGT and its high-level language, which can be offered together as a basic paradigm for expressing and solving complex management and control problems in very different areas. More details and explanations on these and other applications of SGT and programming in SGL can be found in existing technology related publications, like [8–23] and older ones mentioned in Chaps. 1, 3, and 9.

References

1. Spatial operations. ArcMap. https://desktop.arcgis.com/en/arcmap/latest/manage-data/using-sql-with-gdbs/spatial-operations.htm
2. Scholl, M., Voisard, A.: Spatial operations and map operations. In: Liu, L., Özsu, M.T. (eds.) Encyclopedia of Database Systems. Springer, Boston, MA (2009). https://doi.org/10.1007/978-0-387-39940-9_359
3. Olazabal, M., Urzelai, A., et al.: Sustainable spatial management: an integrated approach. ICE Proc. Munic. Eng. **163**(1), 33–41 (2010). https://doi.org/10.1680/muen.2010.163.1.33
4. Mamoulis, N.: Spatial Data Management. Part of the book series: Synthesis Lectures on Data Management (SLDM). Springer (2012). https://doi.org/10.1007/978-3-031-01884-8
5. Gomes, E., da Costa, E.M., Abrantes, P.: Spatial planning and land-use management. Land **13**(1), 94 (2024). https://doi.org/10.3390/land13010094
6. Carazo, M.A.: Distributed operations: the evolution of warfare in the 21st century. ResearchGate, Feb 2005. https://www.researchgate.net/publication/235158726_Distributed_Operations_The_Evolution_of_Warfare_in_the_21st_Century
7. Filipoff, D.: Executing distributed operations in an increasingly contested maritime environment. Report, 1 July 2024. https://www.atlanticcouncil.org/in-depth-research-reports/report/executing-distributed-operations-in-an-increasingly-contested-maritime-environment/
8. Sapaty, P.S.: Spatial management of air and missile defence operation. Math. Mach. Syst. (1) (2023). ISSN 1028-9763. http://www.immsp.kiev.ua/publications/articles/2023/2023_1/01_23_Sapaty.pdf
9. Sapaty, P.S.: A distributed processing system. European Patent N 0389655, Publ. 10.11.93. European Patent Office
10. Sapaty, P.S.: Mobile Processing in Distributed and Open Environments. Wiley, New York (1999)
11. Sapaty, P.S.: Ruling Distributed Dynamic Worlds. Wiley, New York (2005)
12. Sapaty, P.S.: Managing Distributed Dynamic Systems with Spatial Grasp Technology. Springer (2017)
13. Sapaty, P.S.: Holistic Analysis and Management of Distributed Social Systems. Springer (2018)
14. Sapaty, P.S.: Complexity in International Security: A Holistic Spatial Approach. Emerald Publishing (2019)
15. Sapaty, P.S.: Symbiosis of Real and Simulated Worlds under Spatial Grasp Technology. Springer (2021)
16. Sapaty, P.S.: Spatial Grasp as a Model for Space-based Control and Management Systems. CRC Press (2022)
17. Sapaty, P.S.: The Spatial Grasp Model: Applications and Investigations of Distributed Dynamic Worlds. Emerald Publishing (2023)
18. Sapaty, P.S.: Providing Integrity, Awareness, and Consciousness in Distributed Dynamic Systems. CRC Press (2024)
19. Sapaty, P.S.: Spatial Networking in the United Physical, Virtual, and Mental World. Springer (2024)
20. Sapaty, P.S.: Self-Healing and Self-Recovering Systems under the Spatial Grasp Model. Emerald Publishing Limited (2025)
21. Sapaty, P.S.: Global and international security under spatial grasp paradigm. Int. Relat. Dipl. **12**(2), 72–85 (2024). https://doi.org/10.17265/2328-2134/2024.02.004
22. Sapaty, P.S.: Self-recovering-systems-under-spatial-grasp-technology. Cris. Stress. Hum. Resil. Int. J. **6**(2) (2024). https://www.crisisjournal.org/article/126490-self-recovering-systems-under-spatial-grasp-technology
23. Sapaty, P.S.: Spatial grasp model for distributed management and its comparison with traditional algorithms. Int. Relat. Dipl. **13**(3) (2025). https://doi.org/10.17265/2328-2134/2025.03.004. https://www.davidpublisher.com/Public/uploads/Contribute/689ada551780e.pdf

Chapter 7
Spatial Security Applications of SGT

Abstract Spatial security is for protection of systems based on geographical location, and cybersecurity is about protecting systems, networks, and programs from digital attacks. Implementing effective cybersecurity measures is particularly challenging today because there are more devices than people, and attackers are becoming more innovative. After reviewing different spatial security and cybersecurity issues and types of cyberattacks, the Chapter shows in SGL how to find virus sources in distributed networks by tracing infected predecessors or registered infection time in nodes, offers collective malware removing scenario using self-growing and self-organized intelligent cleaning team. It then shows how to organize massive removal of debris around the Earth, how to find and use strongest and weakest network parts, like cliques and articulation points, for improving security of distributed networks, and provides universal network recovery scenario which allows the whole network topology to self-recover after arbitrary damages.

7.1 Introduction

Spatial security refers to the protection of assets and systems based on their geographical location, considering both physical and informational security. It involves understanding location-specific vulnerabilities and implementing appropriate security measures. This can range from securing physical infrastructure like power grids and transportation networks to protecting data and systems that are geographically distributed. ***Space security***, a specific type of spatial security, focuses on the safety and security of space assets and activities, including preventing an arms race in outer space. ***Cybersecurity*** is the combination of methods, processes, tools, and behaviors that protect computer systems, networks, and data from cyberattacks and unauthorized access. ***Spatial cybersecurity*** focuses on protecting space systems, like satellites and ground stations, from cyber threats. This field is crucial due to the increasing reliance of critical infrastructure on space-based services and the growing number of actors in the space industry.

P. S. Sapaty, *Distributed Management with Spatial Grasp Model*, Studies in Systems, Decision and Control 666, https://doi.org/10.1007/978-3-032-19378-0_7

The aim of this Chapter is to review existing spatial security and cybersecurity ideas and concepts and propose concrete and practical solutions of various security management problems with the use of SGT and SGL.

The rest of the Chapter reviews different spatial security and cybersecurity issues and types of cyberattacks; shows how to find virus sources in distributed networks under SGT by the infected predecessors or infection time; offers collective malware removing scenario using self-growing and self-organized intelligent cleaning team in SGL; shows how to use SGT and SGL for massive removal of junk debris around the Earth; describes how to find and use in SGL strongest and weakest network parts like cliques or articulation points for improving security of distributed networks; presents the full network recovery scenario in SGL allowing the whole network to self-recover after arbitrary damages; concludes presentation and cites the used references.

7.2 Spatial Security and Cybersecurity Issues

Reviews different security and cybersecurity solutions and systems with references to the related papers.

Recently, many satellites have been launched for providing global internet broadband service to individual consumers. As ***satellite networks*** become wide-spread, security issues become a central concern, especially the routing protocol of satellite network which guarantees the normal forwarding of network data. Once the routing protocol is attacked, the routing discovery and maintenance will not be completed [1].

Satellite-terrestrial integrated network (STIN) is an indispensable component of the Next Generation Internet (NGI) due to its wide coverage, high flexibility, and seamless communication services. It uses the part of satellite network to provide communication services to the users who cannot communicate directly in terrestrial network. It models the routing problem in satellite network and formulates its combinatorial optimization problem [2].

The LEO satellite network is an important part of the communication network, among which the inter-satellite link (ISL) undertakes the main data transmission task. The routing technology in LEO constellation is an important technical component of the satellite communication system. However, the design of ISL routing algorithm is not easy [3].

Space-air-ground integrated network (SAGIN) is still in nascent stage of design. Despite of several key insights onto the augmentation of terrestrial, aerial and satellite systems, SAGIN ecosystem is yet not mature enough to survive into the realty. More agility, robustness, flexibility, and scalability are required to conform to optimum standard of abstraction [4].

Space information networks is network systems that can receive, transmit, and process spatial information lively. It uses satellites, stratosphere airships, Unmanned Aerial Vehicles, and other platforms as the carrier. It supports high-dynamic, real-time broadband transmission of earth observations and ultra-long-distance, long-delay reliable transmission of deep space exploration [5].

Understanding the ***spatial patterns of urban land use*** at both the macro and the micro levels is a central issue in global change studies. Due to the nonlinear features associated with land use spatial patterns, it is currently necessary to provide some distinct analysis methods to analyze them across a range of remote sensing imagery resolutions [6].

Cybersecurity is the practice of protecting systems, networks, and programs from digital attacks. These are usually aimed at accessing, changing, or destroying sensitive information, or interrupting normal business processes. Implementing effective cybersecurity measures is particularly challenging today because there are more devices than people, and attackers are becoming more innovative [7].

Space Cybersecurity (SC) is becoming critical due to the essential role of space in global critical infrastructure—enabling communication, safe air travel, maritime trade, weather monitoring, environmental surveillance, financial services, and defence systems. Simultaneously, involving diverse stakeholders in space operations further amplifies this criticality [8].

A ***cyber attack*** usually refers to an action designed to target a computer or any element of a computerized information system to change, destroy, or steal data, as well as exploit or harm a network. Cyber attacks have been on the rise, in sync with the digitization of business that has become more and more popular in recent years [9]. Some types are shown in Fig. 7.1.

More extended list of cyberattacks, as follows from [9]: Denial-of-service (DoS) attack, Man-in-the-middle (MITM) attacks, Phishing attack, Whale-phishing attack, Spear phishing, Ransomware, Password attacks. Query Language (SQL) injection, URL interpretation, Domain Name System (DNS) spoofing, Session hijacking,

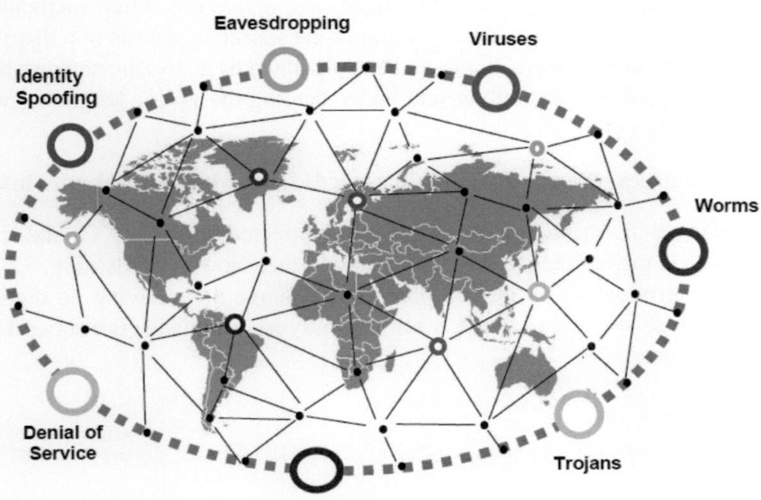

Fig. 7.1 Typical internet cyber attacks

Brute-force attack, Web attacks, Insider threats, Trojan horse, Drive-by attack, XSS attacks, Eavesdropping attacks, Birthday attack, Malware attack.

Global and international security cannot be provided from a single point or a set of separate points whatever powerful these might be. It should rather be deeply embedded and integrated with bodies of real systems wherever in physical, virtual, or combined spaces they may exist. So global security capabilities should not only be distributed, but rather be really spatial, self-organized, and dynamic [10].

The following examples and solutions are using SGT and SGL [10–23] (also their older sources cited in Chaps. 1, 3, and 9).

7.2.1 Fighting Virus Sources in Distributed Networks Under SGT

- **What is a worm virus**

A ***worm virus*** refers to a malicious program that replicates itself, automatically spreading through a network. In this definition of computer worms, the worm virus exploits vulnerabilities in your security software to steal sensitive information, install backdoors that can be used to access the system, corrupt files, and do other kinds of harm. Worms consume large volumes of memory, as well as bandwidth. This results in servers, individual systems, and networks getting overloaded and malfunctioning. A worm is different from a virus, however, because a worm can operate on its own while a virus needs a host computer. To effectively fight viruses in networks, a multi-layered approach combining preventive measures, active protection methods, and proper response procedures is crucial. Finding the source of a virus in a distributed network is a complex problem, but it can be approached by analyzing network traffic, identifying suspicious connections, and understanding the virus's behaviour within the network [24].

- **Finding sources when nodes contain records from which neighbours infected**

The following SGL scenario starting from any infected node (like C) and tracing virus source via the infected predecessors may be as follows, see also Fig. 7.2. This spatial cycle terminating in the believable virus source node having no registered infection predecessor, and its name is subsequently issued outside the network from the starting node.

```
hop_direct(C); STATUS == infected;
output(
  repeat(nonempty(Infected_from); hop(Infected_from)); NAME)
```

- **Finding sources by the infection time registered in nodes**

Starting from any nodes having infection time registered, the scenario moves to the neighboring nodes with earlier infection time, repeating this until such neighbors

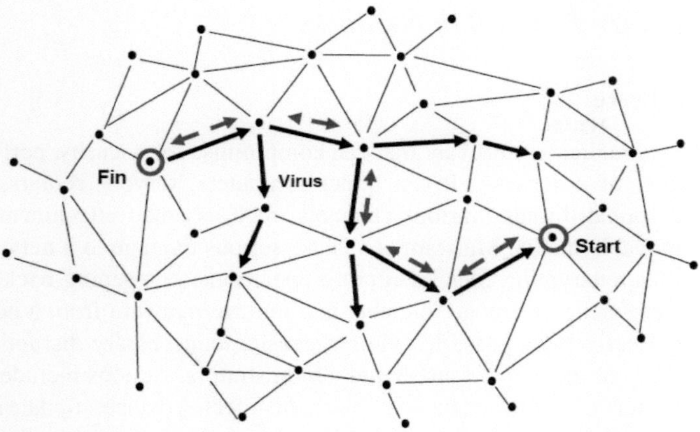

Fig. 7.2 Finding virus source by moving via infected predecessor nodes

exist, as in Fig. 7.3. This spatial process can potentially terminate in more than a single node, and the node with the earliest infection time discovered is proclaimed either the virus source or closest to it (as the latter may be in hiding, i.e. self-protected from discovery).

```
hop_direct(any reachable nodes);
STATUS == infected; frontal(Time) = TIME;
output_min(
  repeat(
    hop(all_links; STATUS == infected; Infection_time < Time;
    Time = Infection_time);
  Time && NAME)
```

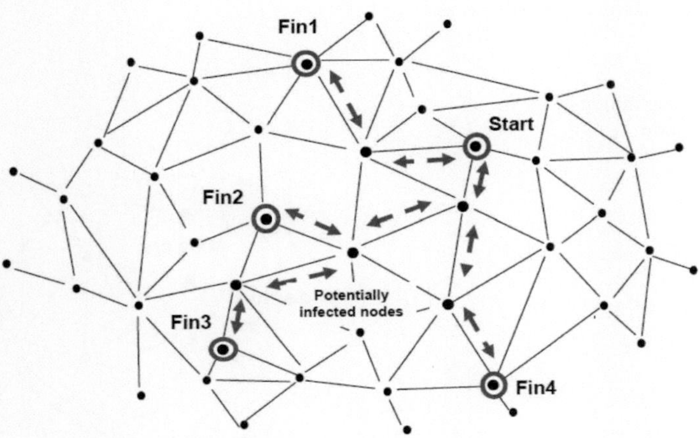

Fig. 7.3 Finding most probable virus source via infection time in nodes

7.2.2 Malware Removal in Networks

- **About malware**

Malware is any malicious software that can compromise the security, performance, or functionality of a network. It can infect computers, servers, routers, or other devices, and spread through various channels, such as email attachments, downloads, or removable media. Malware can cause serious damage to a network, such as stealing data, encrypting files, disrupting operations, or opening backdoors for attackers. Therefore, it is crucial to detect and remove malware from a network as quickly and effectively as possible, without causing unnecessary disruption to the normal activities of the network users and administrators. Actions include: identify the malware, isolate the infected devices, clean the infected devices, update and patch the network, monitor and audit the network, educate and train the network users, and so on [25].

- **Collective malware removal in a distributed network**

The following is a sketch of malware discovery and elimination in distributed networks in SGL by ***self-growing and self-matching spatial virtual team***. The latter can also discover and analyze new and unusual cases and effectively learn and use them in subsequent malware finding-cleaning. The initial stage of this spatial network matching process is symbolically shown in Fig. 7.4.

```
frontal(Latest); nodal(Discovered);
hop_nodes(malware_discovered);
repeat(max)(
  if(malware, (enrich(Discovered, malware_type);
        analyze_clean(Discovered)));
  Latest = Discovered; hop(neighbours);
  if(no(hop(community, BEFORE)), free_linkup(community, BEFORE));
  update(Discovered, Latest))))
```

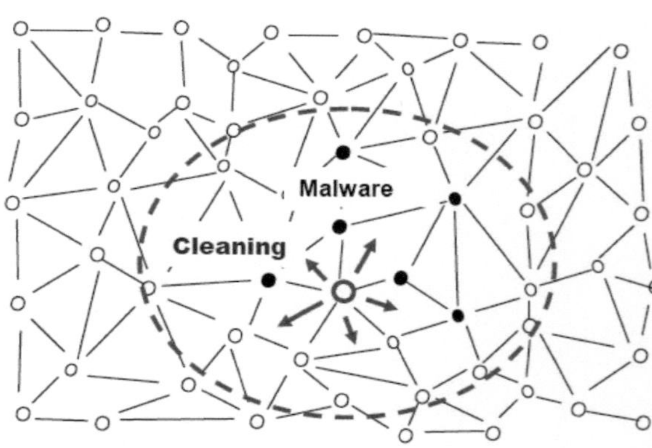

Fig. 7.4 Task definition of cleaning injection

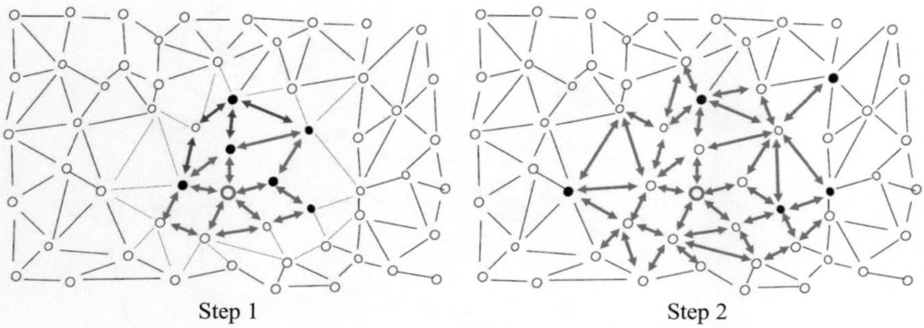

Fig. 7.5 Self-growing-analyzing-optimizing-cleaning spatial community

Such spatially self-growing, self-analyzing, and self-learning team solution (two steps shown in Fig. 7.5) can also self-recover in SGL and operate in case of any possible damages to the network structure (see also Sect. 7.4). All this may be superior to traditional malware discovering and cleaning systems operating individually and separately.

7.3 Removing Multiple Junk Items in Space

- **About junk items**

Removing junk items, or space debris, from orbit involves a combination of strategies, including actively removing larger objects, preventing future debris creation, and relying on natural atmospheric drag to deorbit smaller pieces. Active debris removal (ADR) technologies include robotic capture systems, space tugs, and even lasers to deorbit or dismantle debris. There have been various approaches to managing and mitigating debris, ranging from legislative/regulatory efforts to technological ones [26].

- **Debris removing in SGL**

Different junk items can be found by ground stations launching deorbiting processes simultaneously, as by the following scenario using stations G1 and G2 (as in Fig. 7.6), each trying to find and remove suitable junks independently by the common network of cleaners.

```
hop(G1, G2); IDENTITY = NAME;
frontal(Details) = find_select(junk, TIME);
hop_first(any_cleaner);
repeat(
  Snapshot = parameters(closest_junk, seen);
  if(match(Details, Snapshot),
    (deorbit(Snapshot); abort));
```

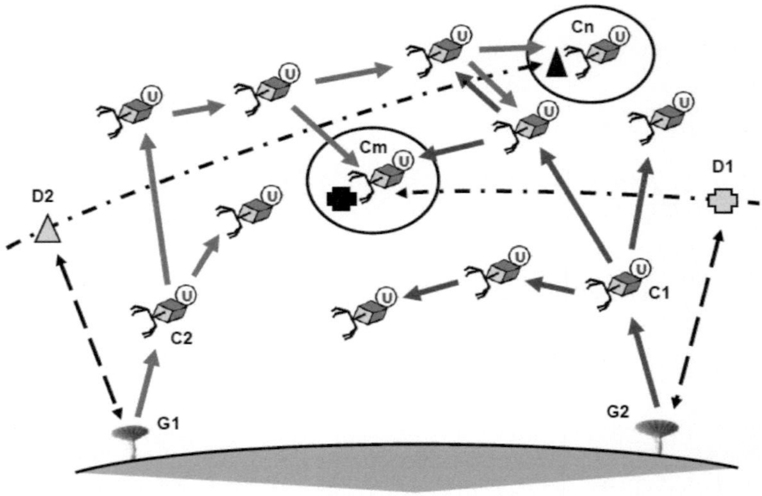

Fig. 7.6 Simultaneous junk reaching and removal

```
update(Details, TIME);
hop_first(other_cleaners, direct_links))
```

7.4 Network Security

7.4.1 Network Security Issues

Network security refers to the measures, technologies, and practices implemented to protect a computer network and its resources from unauthorized access, misuse, or theft. It encompasses various aspects, including firewalls, access control, intrusion detection, and more, all aimed at maintaining the confidentiality, integrity, and availability of network data and systems. Some related ideas and publications as follows.

Network security is an umbrella term to describe security controls, policies, processes and practices adopted to prevent, detect and monitor unauthorized access, misuse, modification, or denial of a computer network and network-accessible resources [27].

Network infrastructure security refers to the strategies, policies, and practices designed to protect an organization's networking components, such as routers, switches, and servers, from various cyber threats and unauthorized access, ensuring the integrity and functionality of connected systems [28].

A ***network disaster recovery plan*** is a set of procedures designed to prepare an organization to respond to an interruption of network services during a natural or humanmade catastrophe. Voice, data, internet access and other network services often share the same network resources [29].

The Internet-of-Things (IoT) has been extensively investigated as an enabling technology to support massive devices or sensors. The ***space information networks*** (SIN) are adopted to provide reliable communications for IoT devices at anytime and anywhere [30].

Space Information Network (SIN) is a multi-purpose heterogeneous network. Due to the large-scale of SIN, its secure and stable operation is vulnerable to various threats. Much of current threat analysis for SIN is based on the network function or architecture [31].

Network security is an umbrella term to describe security controls, policies, processes and practices adopted to prevent, detect and monitor unauthorized access, misuse, modification, or denial of a computer network and network-accessible resources [27].

A ***smurf attack*** is a *distributed denial-of-service attack* in which large numbers of Internet Control Message Protocol (ICMP) packets with the intended victim's spoofed source IP are broadcast to a computer network using an IP broadcast address [32].

7.4.2 Strongest and Weakest Parts of Networks

- **Cliques**

Cliques are traditionally considered strongest parts of networks. The following scenario is finding all full cliques in parallel in the network of Fig. 7.7, with three or more nodes in them (a single four-node clique is shown in bold).

```
hop_nodes(all); frontal(Clique) = NAME;
repeat(hop_links(all); not_belong(NAME, Clique);
      yes(and_parallel(hop(links(any), nodes(Clique)));
      if(PREDECESSOR > NAME, append(Clique, NAME), blind));
count(Clique) >= 3; output(Clique)
```

The following cliques will be issued in the nodes where their collection terminated:

```
(B, D, G, Z), (B, C, Y), (C, E, F)
```

Fig. 7.7 Strongest parts, or cliques, in the network

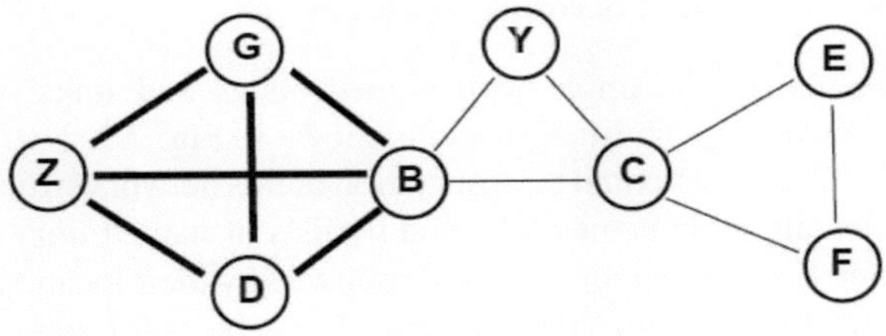

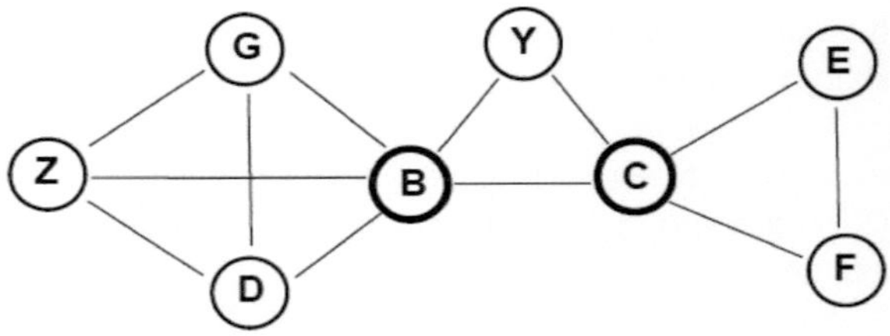

Fig. 7.8 Weakest or articulation points of the network

- **Articulation points**

These are traditionally considered as weakest pats of the network, like nodes `F` and `H` in Fig. 7.8. Each of them, when removed, splits the network into disjoint parts. Below is parallel and fully distributed solution for finding all articulation points in the whole network.

```
hop_nodes(all); IDENTITY = NAME;
hopfirst_node(current);
stay(hopfirst_random(links(all));
      repeat(hopfirst_link(all)));
if(hopfirst_links(all), output(NAME))
```

Names of articulation points `B` and `C` will be issued by these nodes themselves.

Finding and investigating these strongest and weakest features of distributed networks can be important and even crucial for their security and stability. For example, cliques, as fully organized collectives with certain types of nodes and relations between them, can be either very useful for system-strengthening or dangerous under certain circumstances, where reducing their connections with the rest of the network or even fully destroying them may be the solutions needed. On the opposite, articulation points may need additional linking them with other network nodes (or connecting other nodes with each other) to strengthen the network as a whole and preserve its integrity. Such network modifications can be effectively expressed and implemented in SGL under SGT, which was also discussed in the previous publications.

7.4.3 *Full Network Recovery*

- **Distributed network representation**

A network example with named nodes and links, which is distributed in two-dimensional physical space, is shown in Fig. 7.9.

Compact textual representation of this network structure assigned to variables Top (having node names followed by lists of named links leading to named neighbors) and Loc (with node names w followed by their location coordinates #w expressed in some physical or virtual notation) may be as follows.

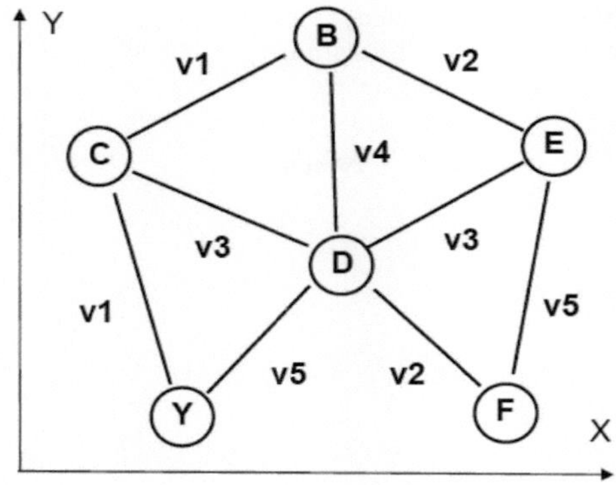

Fig. 7.9 Representation of distributed network topology

```
Top = (B:(v1:C, v2:E, v4:D), C:(v1:B, v3:D, v1:Y), E:(v2:B, v3:D,
v5:F),
       D:(v3:C, v4:B, v3:E, v5:Y, v2:F), Y:(v1:C, v5:D), F:(v2:D,
v5:E));
Loc = (B:#B, C:#C, E:#E, D:#D, Y:#Y, F:#F)
```

Such network description may be very useful for the network self-recovery after damages to its structure and loss of any nodes and links, even if a single node still remains alive.

- **Network self-healing in SGL**

We can easily write in SGL a full network recovery scenario which, initially starting in all still existing nodes, will allow the whole topology self-recovery after arbitrary damages. Each node, seeing absence of its registered neighbors, will be providing their restoration, including supplying the recovered nodes with the overall network data in Top and Loc. This may continue in parallel spanning tree mode to the full depth, but if the expected neighbor already exists, it stops network coverage (with restoring link to it if absent). The following scenario starts automatically and simultaneously in all nodes which still exist (presuming each already having full Top and Loc network copies set up in them earlier, before the network damage).

```
hop_nodes(all); Topf = Top; Locf = Loc;
repeat(
  split(Topf:NAME);
  or_seq((empty(Locf:VAL[2]);
      create(link(VAL[1], node(VAL[2], coord(Locf:VAL[2]))));
      Top = Topf; Loc = Locf),
      (no(hop(link(VAL[1], node(VAL[2]));
      quit_linkup(VAL[1], node(VAL[2]))))
```

Any number of nodes may happen to be destroyed, and then fully recovered by starting in still existing nodes (like Y and E in Fig. 7.10).

More on self-recovery of distributed networks in SGL can be found in [22, 23].

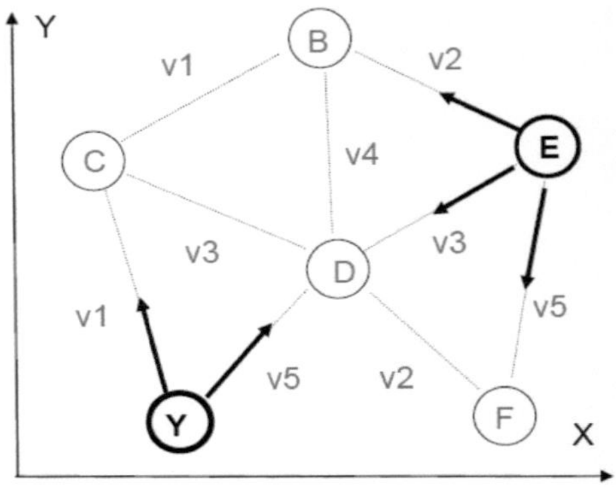

Fig. 7.10 Distributed system self-recovery starting from the remaining nodes

7.5 Conclusions

After reviewing different spatial security and cybersecurity issues and types of cyberattacks, the Chapter investigated and offered spatial management solutions for solving complex security problems in very different areas, ranging from terrestrial to celestial. It also showed how to investigate, improve, and repair large distributed networked systems after any damages, including their complete self recovery. Similar to the conclusion of Chap. 6, SGT and SGL can be considered together as a basic paradigm for investigating and solving complex management and control problems, this time in the areas of security and cybersecurity too.

References

1. Yan, Y., Han, G., Xu, H.: A survey on secure routing protocols for satellite network. J. Netw. Comput. Appl. **145**, 102415 (2019). https://www.sciencedirect.com/science/article/abs/pii/S1084804519302498
2. Yin, Y., Huang, C., et al.: Reinforcement learning-based routing algorithm in satellite-terrestrial integrated networks. Hindawi Wirel. Commun. Mob. Comput. **2021**, Article ID 3759631. https://doi.org/10.1155/2021/3759631
3. Zhao, N., Long, X., Wang, J.: A multi-constraint optimal routing algorithm in LEO satellite networks. Wireless Netw. (2021). https://doi.org/10.1007/s11276-021-02674-3
4. Ray, P.P.: A review on 6G for space-air-ground integrated network: key enablers, open challenges, and future direction. J. King Saud Univ. Comput. Inf. Sci. **34**(9) (2022). https://doi.org/10.1016/j.jksuci.2021.08.014
5. Zhuo, M., Liu, L., et al.: Survey on security issues of routing and anomaly detection for space information networks. Sci. Rep. https://doi.org/10.1038/s41598-021-01638-z
6. Wu, H., Sun, Y., et al.: Examining the satellite-detected urban land use spatial patterns using multidimensional fractal dimension indices. Remote Sens. **5**(10) (2013). https://doi.org/10.3390/rs5105152
7. What is cybersecurity? https://www.cisco.com/site/us/en/learn/topics/security/what-is-cybersecurity.html

8. Khan, S.K., Shiwakoti, N., et al.: Space cybersecurity challenges, mitigation techniques, anticipated readiness, and future directions. Int. J. Crit. Infrastruct. Prot. **47**, 100724 (2024). https://doi.org/10.1016/j.ijcip.2024.100724
9. Types of cyber attacks. https://www.fortinet.com/uk/resources/cyberglossary/types-of-cyber-attacks
10. Sapaty, P.S.: Global and international security under spatial grasp paradigm. Int. Relat. Dipl. **12**(2), 72–85 (2024). https://doi.org/10.17265/2328-2134/2024.02.004
11. Sapaty, P.S.: A distributed processing system. European Patent N 0389655, Publ. 10.11.93. European Patent Office
12. Sapaty, P.S.: Mobile Processing in Distributed and Open Environments. Wiley, New York (1999)
13. Sapaty, P.S.: Ruling Distributed Dynamic Worlds. Wiley, New York (2005)
14. Sapaty, P.S.: Managing Distributed Dynamic Systems with Spatial Grasp Technology. Springer (2017)
15. Sapaty, P.S.: Holistic Analysis and Management of Distributed Social Systems. Springer (2018)
16. Sapaty, P.S.: Complexity in International Security: A Holistic Spatial Approach. Emerald Publishing (2019)
17. Sapaty, P.S.: Symbiosis of Real and Simulated Worlds under Spatial Grasp Technology. Springer (2021)
18. Sapaty, P.S.: Spatial Grasp as a Model for Space-based Control and Management Systems. CRC Press (2022)
19. Sapaty, P.S.: The Spatial Grasp Model: Applications and Investigations of Distributed Dynamic Worlds. Emerald Publishing (2023)
20. Sapaty, P.S.: Providing Integrity, Awareness, and Consciousness in Distributed Dynamic Systems. CRC Press (2024)
21. Sapaty, P.S.: Spatial Networking in the United Physical, Virtual, and Mental World. Springer (2024)
22. Sapaty, P.S.: Self-Healing and Self-Recovering Systems under the Spatial Grasp Model. Emerald Publishing Limited (2025)
23. Sapaty, P.S.: Self-recovering-systems-under-spatial-grasp-technology. Cris. Stress. Hum. Resil. Int. J. **6**(2) (2024). https://www.crisisjournal.org/article/126490-self-recovering-systems-under-spatial-grasp-technology
24. Worm virus definition. https://www.fortinet.com/uk/resources/cyberglossary/worm-virus#:~:text=A%20worm%20virus%20refers%20to,virus%20needs%20a%20host%20computer
25. How can you remove malware from a network without disruption? Last updated on 29 May 2025. https://www.linkedin.com/advice/0/how-can-you-remove-malware-from-network-without-dwj2c
26. Iyer, V.: How do you clean up 170 million pieces of space junk? 05.24.23. https://fas.org/publication/how-do-you-clean-up-170-million-pieces-of-space-junk/
27. Network security. https://en.wikipedia.org/wiki/Network_security#cite_note-1
28. Network infrastructure security—best practices and strategies. https://www.dataguard.com/blog/network-infrastructure-security-best-practices-and-strategies/
29. Peterson, S., Hilliard, J.: What is a network disaster recovery plan? TechTarget. Published: 14 Dec 2022. https://www.techtarget.com/searchdisasterrecovery/definition/Network-disaster-recovery-plan
30. Liu, J., Bai, L., Jiang, C., Zhang, W.: Space network security. In: Space-Air-Ground Integrated Network Security. Springer, Singapore (2023). https://doi.org/10.1007/978-981-99-1125-7_5
31. Wu, X., Du, Y., Fan, T., et al.: Threat analysis for space information network based on network security attributes: a review. Complex Intell. Syst. **9**, 3429–3468 (2023). https://doi.org/10.1007/s40747-022-00899-z
32. Smurf attack. https://en.wikipedia.org/wiki/Smurf_attack

Chapter 8
Spatial Grasp Model Versus the Concept of Algorithm

Abstract Algorithm is usually considered as a sequence of rigorous instructions for solving a class of specific problems. A review of definitions of algorithm and its modifications like parallel, distributed, spatial analysis, mobile agents too, is provided. The Chapter compares SGT principles with algorithms, where self-evolving spatial scenarios in SGL often resemble massive spatial phenomena like virus, smog, radiation, rain, smell, light, or opinion, which are far more general and global than algorithmic sequences. It then treats SGT as a self-evolving recursive pattern spatially matching networked systems, potentially any continuous spaces too, up to the whole world. And finally shows how dynamic spatial capabilities of SGT may relate to non-algorithmic ideas especially popular in relation to living organisms and societies.

8.1 Introduction

An ***algorithm*** is a finite sequence of well-defined, unambiguous instructions designed to solve a specific problem or accomplish a particular task. It acts as a step-by-step procedure that takes an input, processes it through a series of operations, and produces a desired output. Algorithms are fundamental to computer science and programming, forming the basis of how computers perform calculations, process data, and make decisions. However, the concept extends beyond computing, as any replicable, step-by-step process to achieve a goal can be considered an algorithm too.

The aim of this Chapter is to review existing definitions of algorithms and their types, compare SGT with the concept of algorithm, and investigate relation of SGT and SGL to non-algorithmic ideas.

The rest of the Chapter reviews algorithms and their existing types, like traditional algorithms, parallel algorithms, distributed algorithms, mobile agents, and spatial analysis algorithms; compares SGT with algorithms from one side, and with massive spatial processes and effects like virus, smog, radiation, rain, smell, light, or opinion, from the other side; exhibits SGT as self-evolving recursive spatial pattern capable of impacting large distributed spaces which may be continuous, not only discrete;

P. S. Sapaty, *Distributed Management with Spatial Grasp Model*, Studies in Systems, Decision and Control 666, https://doi.org/10.1007/978-3-032-19378-0_8

investigates relation of SGT to non-algorithmic ideas becoming popular especially in relation to living systems; concludes presentation and cites the used publications.

8.2 Algorithms, Their Types, and SGT

8.2.1 Traditional Algorithms

Algorithm is a well-established, step-by-step procedure designed to solve a specific problem or perform a computation. These algorithms are typically rule-based, deterministic, and focus on precise execution of steps. Key characteristics of traditional algorithms: step-by-step, deterministic, rule-based, oriented on specific tasks and may not be easily adaptable to other tasks. Also: An algorithm is a finite sequence of well-defined, unambiguous instructions designed to solve a specific problem or accomplish a particular task. It takes an input, processes it through a series of logical steps, and produces a desired output. Algorithms are fundamental to computer science and are used in various fields, from simple tasks like following a recipe to complex computations in machine learning and data analysis. They provide a systematic and reproducible method for achieving a specific goal. Some issues on usual algorithms, expressed by existing publications, may be as follows.

In elementary arithmetic, a standard algorithm or method is a specific method of computation which is conventionally taught for solving particular mathematical problems. These methods vary somewhat by nation and time, but generally include exchanging, regrouping, long division, and long multiplication using a standard notation, and standard formulas for average, area, and volume [1].

Algorithms are step-by-step procedures used to solve problems. There are different variations for recording the steps for each operation (addition, subtraction, multiplication and division). The emphasis in learning must be on understanding. To help develop understanding, connections should be made between the algorithm and place value, base-ten concepts and number properties [2].

In mathematics and computer science, an algorithm is a finite sequence of mathematically rigorous instructions, typically used to solve a class of specific problems or to perform a computation. Algorithms are used as specifications for performing calculations and data processing. More advanced algorithms can use conditionals to divert the code execution through various routes and deduce valid inferences [3].

A procedure for solving a mathematical problem (as of finding the greatest common divisor) in a finite number of steps that frequently involves repetition of an operation. Broadly: a step-by-step procedure for solving a problem or accomplishing some end. Sometimes you solve a problem by coming up with an algorithm of some kind. But sometimes you solve a problem in a very ad hoc sort of way [4].

An algorithm is a procedure used for solving a problem or performing a computation. Algorithms act as an exact list of instructions that conduct specified actions step by step in either hardware- or software-based routines. Algorithms are widely used throughout all areas of IT. In mathematics, computer programming and computer science, an algorithm usually refers to a small procedure that solves a recurrent problem [5].

Algorithm is a set of instructions that is designed to accomplish a task. Algorithms usually take one or more inputs, run them systematically through a series of steps, and provide one or more outputs. Algorithms are typically associated with computing and are an essential element of computer programming. Algorithms can be used to accomplish a variety of computational tasks, such as performing calculations or finding information in databases [6].

An algorithm is a set of steps for accomplishing a task or solving a problem. Typically, algorithms are executed by computers, but we also rely on algorithms in our daily lives. Each time we follow a particular step-by-step process, we are in fact following an algorithm. In the context of computer science, an algorithm is a mathematical process for solving a problem using a finite number of steps [7].

In today's digital age, algorithms are the invisible architects shaping the landscape of technology and information. These complex sets of rules and instructions underpin almost every aspect of our digital lives, from the way search engines prioritize results to how social media feeds present content. At its core, an algorithm is a methodical, step-by-step procedure for solving problems or accomplishing tasks [8].

Elementary algorithm structures are shown in Fig. 8.1, with 8.1a reflecting the flowchart with operations just following each other, and 8.1b exhibiting a cyclic one with decision makers (as 1, 2, 3), like in finding the greatest common divisor of two numbers [3].

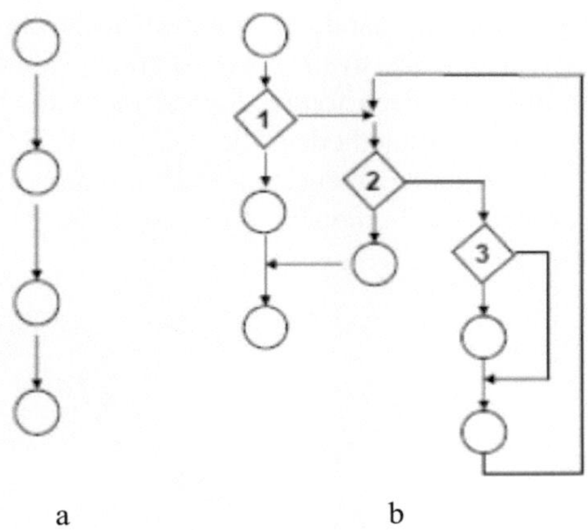

Fig. 8.1 Traditional algorithms: **a** as elementary sequence of instructions, **b** using branching and repetition

8.2.2 Parallel Algorithms

Parallel algorithms are a class of algorithms designed to be executed on multiple processors or computing units simultaneously, in contrast to traditional sequential algorithms that perform operations in a step-by-step, chronological order. The core concept behind parallel algorithms is to divide a larger problem into smaller, independent sub-problems that can be solved concurrently, thereby reducing the overall execution time and improving efficiency, especially for computationally intensive tasks. Key characteristics and concepts of parallel algorithms include: Concurrency, Divide and Conquer, and Parallel Computing Models. Also, parallel algorithms are computational methods designed to perform multiple operations simultaneously, leveraging the capabilities of parallel computing systems. Unlike traditional sequential algorithms that execute instructions one after another, parallel algorithms divide a problem into smaller, independent sub-problems that can be processed concurrently by multiple processors or cores (as in Fig. 8.2).

Some ideas and publications on parallel algorithms, as follows.

In computer science, a parallel algorithm, as opposed to a traditional serial algorithm, is an algorithm which can do multiple operations in a given time. Many parallel algorithms are executed concurrently—though in general concurrent algorithms are a distinct concept—and thus these concepts are often conflated, with which aspect of an algorithm is parallel and which is concurrent not being clearly distinguished [9].

Parallel algorithms refer to a computational approach where multiple solution points or populations of solutions are utilized at each iteration of the algorithm. By employing parallel processors, these algorithms significantly enhance the speed of convergence to the final solution. Additionally, certain types of parallel algorithms have been found to yield more stable and less risky solutions compared to conventional methods [10].

In computer science, analysis of parallel algorithms is the process of finding the computational complexity of algorithms executed in parallel—the amount of time,

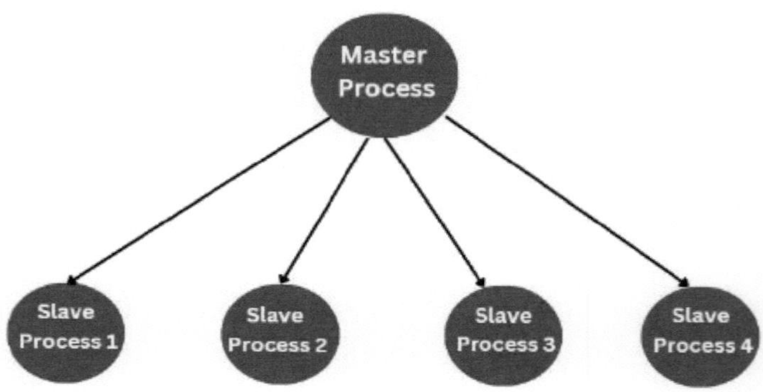

Fig. 8.2 Parallel algorithm

storage, or other resources. In many respects, analysis of parallel algorithms is similar to the analysis of sequential algorithms, but is generally more involved because one must reason about the behaviour of multiple cooperating threads of execution [11].

Parallel computing is the process of distributing a larger task into a small number of independent tasks and then using multiple processing elements simultaneously. Parallel computing is more efficient than the serial approach as it requires less computation time. The need for a parallel algorithm model arises in order to understand the strategy used for the partitioning of data and the ways in which these data are being processed [12].

8.2.3 Distributed Algorithms

Distributed algorithms are algorithms designed to run on multiple, interconnected computational entities (processors or nodes) that do not share a common memory and communicate by message passing. Unlike traditional sequential algorithms that execute on a single machine, distributed algorithms coordinate the actions of these independent entities to achieve a common goal. Key characteristics and aspects of distributed algorithms include: concurrency, coordination, fault tolerance, limited information. Also: distributed algorithms are algorithms designed to run on multiple, interconnected processors or nodes that do not share a common memory and communicate primarily through message passing (see Fig. 8.3).

Some ideas and publications on distributed algorithms, as follows.

A distributed algorithm is the one designed to run on computer hardware constructed from interconnected processors. They are used in different application such as telecommunications, scientific computing, distributed information processing, and real-time process control. Standard problems solved by them include leader election, distributed search, spanning tree generation, and resource allocation [13].

A distributed algorithm is defined as a set of processors running similar programs in parallel, where each processor is represented by an index. The algorithm involves processes that work together simultaneously to achieve a common goal. Compared to centralized algorithms, distributed algorithms reduce communication energy and detection accuracy while increase the processing energy [14].

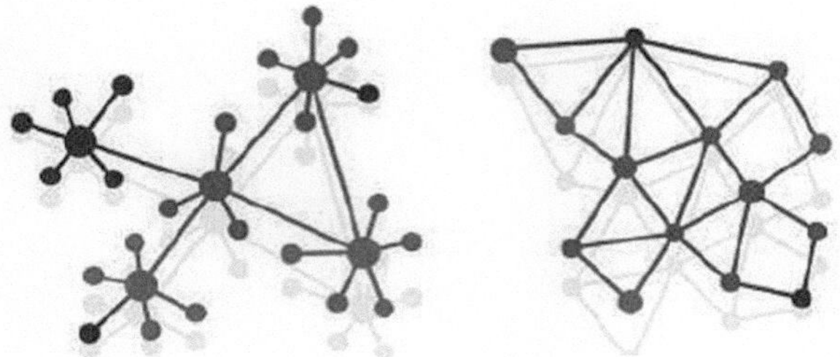

Fig. 8.3 Distributed algorithms

The main goal of Guide to Distributed Algorithms is to provide a detailed study of the design and analysis methods of distributed algorithms and to supply the implementations of most of the presented algorithms in Python language, which is the unique feature of the book not found in any other contemporary books on distributed computing [15].

8.2.4 Mobile Agents

Mobile agents are software programs that can autonomously move and execute across different computer systems in a network. They combine software with data, and are capable of migrating, saving their state, and resuming execution on the destination. This makes them useful for distributed computing, especially in environments with latency, bandwidth limitations, or potential disconnections. Here's a more detailed breakdown: autonomy, mobility, distributed computing, reduced latency and bandwidth resilience. Also: a mobile agent is a software agent that has the ability to move between different computers on a network, continuing its work in a new location (see Fig. 8.4).

Unlike traditional client–server models, where the client requests data and performs actions, a mobile agent moves to the server to perform tasks, thereby delegating work to the server. The main characteristics of mobile agents are: autonomy, mobility, sociality, learning. Some mobile agents can change their behavior based on experience gained during their work. Some mobile agent ideas and publications, as follows.

In computer science, a mobile agent is a piece of software agent combined with data that is able to migrate from one computer to another autonomously and continue its execution on the destination with the ability to interact with other agents there. Rather than a client requesting data and performing actions, a mobile agent is sent to a server to perform those tasks. This paradigm delegates the work from the client and onto the server [16].

Fig. 8.4 Mobile agents

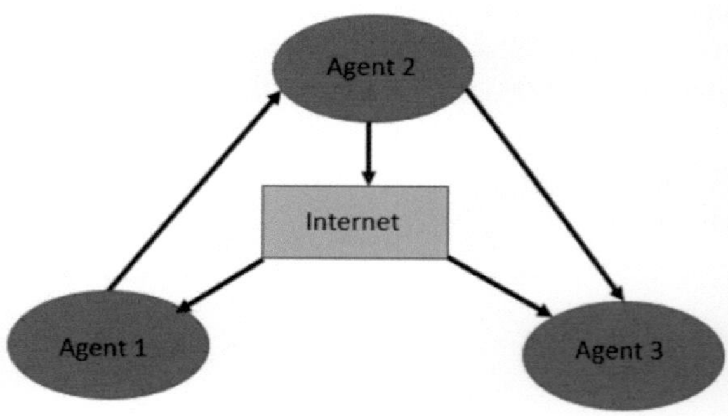

Discusses the potential uses of mobile agents in network management and defines software agents and a navigation model that determines agent mobility. Lists a number of potential advantages and disadvantages of mobile agents and includes a short commentary on the ongoing standardization activity. Comprises descriptions of several actual and potential applications of mobile agents [17].

A mobile agent is an object that migrates through many nodes of a heterogeneous network of computers, under its own control, in order to perform tasks using resources of these nodes. The use of this technology represents a change in the distributed programming paradigm. This approach provides many benefits to the development of distributed applications but introduces new requirements to the engineering of these systems [18].

8.2.5 Spatial Analysis Algorithms

Spatial analysis algorithms are computational methods used to analyze geographic data, uncovering patterns, relationships, and trends within spatial distributions. These algorithms are essential for tasks ranging from mapping and navigation to environmental modelling and urban planning. Key types of spatial analysis algorithms: Geometric Algorithms: Statistical Spatial Analysis: Spatial Indexing: Network Analysis, Urban Planning, Environmental Monitoring, Geospatial Engineering. Also: Spatial analysis algorithms are techniques used to analyze and understand geographic data by examining spatial relationships and patterns. These algorithms are crucial for various fields, including geography, urban planning, and environmental science, enabling the extraction of meaningful insights from spatial data (see Fig. 8.5).

Some ideas and publications on spatial analysis algorithms, as follows.

Spatial analysis is any of the formal techniques which study entities using their topological, geometric, or geographic properties, primarily used in urban design.

Fig. 8.5 Spatial analysis algorithms

Spatial analysis includes a variety of techniques using different analytic approaches, especially spatial statistics. It may be applied in fields as diverse as astronomy, with its studies of the placement of galaxies in the cosmos, or to chip fabrication engineering [19].

Geospatial algorithms are key tools in Geospatial Engineering, helping to analyze and manage spatial data effectively. They enhance tasks like mapping, navigation, and environmental modelling. Include: spatial indexing, shortest path algorithms, map projection algorithms, spatial clustering algorithms, terrain analysis algorithms, coordinate transformation algorithms, and others [20].

Algorithm visualization is a widely-used tool for teaching data structures and algorithms. Spatial data algorithms are algorithms that are designed to process multidimensional data. This introduces a spatial data extension to the successful learning environment, which includes automatically assessed visual algorithm simulation exercises. First impressions on using the visualizations in teaching are also described [21].

8.2.6 From Algorithms to SGT

Spatial Grasp Technology with its basic Spatial Grasp Language ([22–37], older sources cited in Chaps. 1, 3, and 9) effectively covers existing algorithmic ideas and their expressions and implementations, which were practically conformed in numerous solutions of the existing publications. Direct relation to algorithmic ideology is also in different SGL scenarios of the previous chapters of this book, as named below.

Traditional algorithms: full network creation (Chap. 5).

Distributed and parallel algorithms: creating and using shortest path tree, spatial pattern matching (Chap. 6); removing multiple junk items in space, distributed system self-recovery (Chap. 7).

Mobile agents: finding individuals worldwide, tracking mobile objects (Chap. 6); fighting virus sources in distributed networks, removing multiple junk items in space (Chap. 7).

Spatial analysis algorithms: converting a curved way to the sequence of straight paths, finding perimeters of complex regions, establishing network security borders (Chap. 5); cooperative finding of oil spill center (Chap. 6).

8.3 SGT and Massive Spatial Phenomena

SGT philosophy, behaviour, and capability often relate to and resemble massive and global processes and features which may directly or indirectly relate to such phenomena like virus, smog. radiation, rain, smell, light, opinion, fear, and others, with some explanations and publications on these as follows. These features are often very different and much broader than of the concept of algorithm.

- **Examples of natural spatial processes and effects**

A ***virus*** is a submicroscopic infectious agent that replicates only inside the living cells of an organism. Viruses infect all life forms, from animals and plants to microorganisms, including bacteria and archaea. Viruses are found in almost every ecosystem on Earth and are the most numerous type of biological entity [38].

Smog is the chemical reaction of sunlight, nitrogen oxides, and volatile organic compounds in the atmosphere, which leaves airborne particles and ground-level ozone. Ground level ozone is not emitted directly into the air, but is created by chemical reactions in the presence of sunlight [39].

Radiation is energy. It can come from unstable atoms that undergo radioactive decay, or it can be produced by machines. Radiation travels from its source in the form of energy waves or energized particles. There are different forms of radiation and they have different properties and effects [40].

A ***flood*** is an overflow of water (or rarely other fluids) that submerges land that is usually dry. In the sense of "flowing water", the word may also be applied to the inflow of the tide. Floods are of significant concern in agriculture, civil engineering and public health [41].

Rain is a form of precipitation where water droplets that fall under gravity. Rain is a major component of the water cycle and is responsible for depositing most of the fresh water on the Earth. The major cause of rain production is moisture moving along three-dimensional zones of temperature [42].

The sense of ***smell*** is the special sense through which smells (or odors) are perceived. The sense of smell has many functions, including detecting desirable foods, hazards, and pheromones, and plays a role in taste. In humans, it occurs when an odor binds to a receptor within the nasal cavity [43].

Light, otherwise known as electromagnetic radiation, is different from anything else in the universe. Light is not matter, and it is not made up of pieces of matter. Since it is different, we must think of ways to explain light. We consider two analogies to things we know. We model light as a wave and as a particle [44].

Public ***opinion***, or popular opinion, is the collective opinion on a specific topic or voting intention relevant to society. It is the people's views on matters affecting them. Public opinion is heavily influenced by the media, many studies have been undertaken which look at the different factors which influence opinion [45].

Fear is an unpleasant emotion that arises in response to perceived dangers or threats. Fear causes physiological and psychological changes. Extreme cases of fear can trigger an immobilized freeze response. Fear in humans can occur in response to a present stimulus or anticipation of a future threat [46].

- **Similarity with massive spatial coverage of SGL scenarios**

Many complex solutions in large distributed systems provided by SGT and expressed in SFL strongly resembled spatial features of the mentioned above massive natural processes and phenomena. These can be found in the cited previous publications on SGT and also in the previous chapters of this book named as follows:

Examples of massive spatial coverage in SGL: finding the network dimensions and surrounding borders creation, finding the network area centre and creating circular border, directly linking existing nodes within certain distance from each other, finding perimeters of complex regions (Chap. 5); finding individuals worldwide, spatial pattern matching, battling forest fires (Chap. 6); finding strongest and weakest parts of networks (Chap. 7).

8.4 SGT as Self-Evolving Spatial Pattern

SGT as a universal spatial paradigm may have many faces, bodies, treatments or explanations, and its particular pattern features are explained below.

A ***recursive spatial grasp pattern***, within the SGT framework ([22–37], other references in Chaps. 1, 3, 9) refers to a self-evolving organized substance that propagates through a distributed space while replicating, modifying, managing, and matching the environment. These patterns, expressed in SGL, are inherently parallel and distributed, enabling the creation of active spatial infrastructures that can adapt to various systems and models. Their main features include recursion, self-evolution, self-matching, self-propagation, self-replication and modification, self-control and decision making, holism, adaptability, flexibility, and even consciousness. They are often behaving like ***spatial organisms***.

An example of a pattern in a discrete networking form is shown in Fig. 8.6.

The result of its self-matching with a distributed network, as in Fig. 8.7, is symbolically shown below in SGL.

```
hop(all); frontal(F0, F1, F2); F0 = ADDR; pass(m1); F1 = ADDR;
pass(m8);
and(
  (pass(m9); F2 = ADDR; hop(m4, F1); pass(m2); hop(m3, F2); hop(m11,
F0)),
  (pass(m10); F2 = ADDR; hop(m7, F1)); pass(m5); hop(m6, F2);
hop(m12, F0)))
```

This self-matching may also provide certain network actions or changes if needed, say, removing all matched nodes with their links to other nodes, as follows.

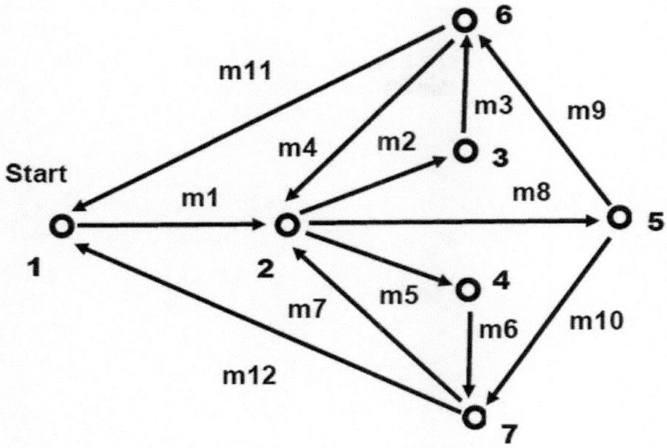

Fig. 8.6 Discrete spatial pattern

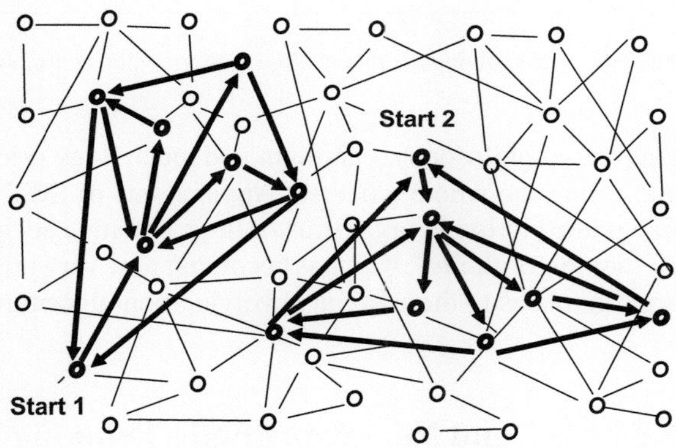

Fig. 8.7 Spatial pattern matching in discrete environment

```
hop(all);
if(
  (frontal(F0, F1, F2); F0 = ADDR; pass(m1); F1 = ADDR; pass(m8);
      and(
       (pass(m9); F2 = ADDR; hop(m4,F1); pass(m2); hop(m3,F2);
hop(m11,F0)),
       (pass(m10); F2 = ADDR; hop(m7,F1)); pass(m5); hop(m6,F2);
hop(m12,F0)))),
  remove(HISTORY, all_nodes))
```

We considered only discrete network-based pattern representation and the environment it was matched with. But SGL (ideologically as well as technically) allows for a much broader and universal problem definitions and formulations, where spatial pattern-formulated tasks can be continuous, recursive, growing, self-evolving, actually ***alive***, impacting the whole continuous world (as symbolically shown in Fig. 8.8).

In comparison with the described SG Model and Language, the traditional and extended algorithms may be considered just as specific tools for solving concrete problems. Whereas the Spatial Grasp can actually, not only symbolically, represent

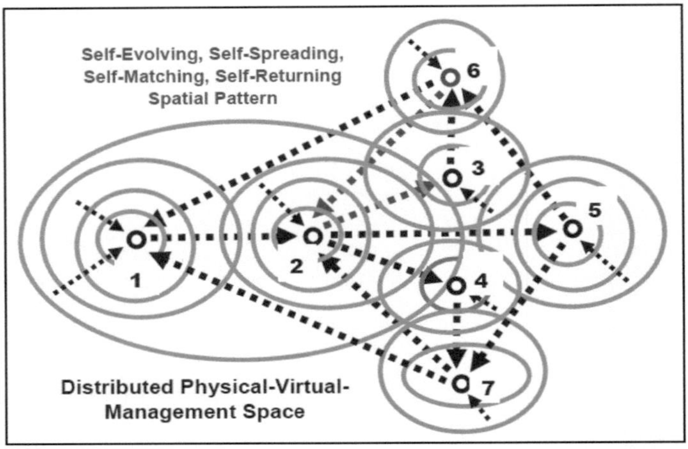

Fig. 8.8 Self-evolving continuous SGL pattern in the continuous distributed space

a universal philosophy, paradigm and technology oriented on explaining, covering, and ruling the whole universe. Applications of SGT to some classes of problems, like related to living organisms or large dynamic societies, may even be considered as "non-algorithmic", which is becoming now very popular and worldwide. The next section relates to these rapidly evolving non-algorithmic ideas and thinking.

8.5 SGT and Non-algorithmic Thinking

Non-algorithmic refers to processes or systems that do not follow a predetermined, step-by-step procedure, unlike algorithms. This often implies a more flexible, adaptive or even unpredictable behaviour, which can be contrasted with the deterministic nature of algorithms. Non-algorithmic methods often rely on heuristics, intuition, experience, or other unstructured ways of making decisions. With its integral, dynamic, recursive, and self-organized spatial philosophy and implementation, ***fully covering the concept of algorithm*** and its extensions, ***SGT is also may be close to non-algorithmic ideas and thinking***, and can effectively solve different problems from the related areas, including the following.

Algorithms work blind, ignoring the meanings of involved symbols and the aims of their actions. They are stated in a fixed language; hence, they are language dependent. Introduces procedures identifying entities by their attributes, as the scientific method requires, and terms them eulerithms. These are language-independent procedures being able to build algorithms [47].

It argues that in standard non-algorithmic contexts, such as hirings, we do not think that lack of calibration entails unfair bias, and that it is difficult to see why algorithmic contexts, as it were, should differ fairness-wise from non-algorithmic ones in this respect. Hence, we should reject the view that calibration is necessary for fairness in an algorithmic context [48].

Understanding the biological organization from non-algorithmic mathematical models is one of the most important fields in mathematical biology. However, biological thinking in this century witnessed the overwhelming use of the algorithmic and syntactic mathematics that cannot deal with the autopoietic self-referential organization of living systems [49].

In the context of critical thinking, "non-algorithmic" refers to the aspects of thought and problem-solving that are not easily captured by a set of predefined, step-by-step instructions or rules (algorithms). It encompasses the ability to analyze information, evaluate evidence, make reasoned judgments, and solve complex problems in ways that may not be easily automated or predicted by a computer program [50].

The Emperor's New Mind: Concerning Computers, Minds and The Laws of Physics is a 1989 book by the mathematical physicist Roger Penrose. He argues that human consciousness is non-algorithmic, and thus is not capable of being modelled by a conventional Turing machine, which includes a digital computer. Penrose hypothesizes that quantum mechanics plays an essential role in the understanding of human consciousness [51].

In his The Emperor's New Mind Roger Penrose propounds and defends the thesis that human cognition is beyond the algorithmic. Since "quantum computing," when rigorously unpacked, doesn't provide a formal foundation for the Penrosian position, and since the same is true of the connectionist's analog "neural computing," is there perhaps reason to think the Penrosian position simply cannot be founded upon formal results [52].

The existence of a non-algorithmic side of the mind, conjectured by Penrose on the basis of Gödel's first incompleteness theorem, is investigated here in terms of a quantum metalanguage. It suggests that, besides human ordinary thought, which can be formalized in a computable, logical language, there is another important kind of human thought, which is Turing-noncomputable. This is metathought, the process of thinking about ordinary thought [53].

One of the most compelling problems in science consists in understanding how living systems process information. After all, the way they process information defines their capacities to learning and adaptation. There is an increasing consensus in that living systems are not machines in any sense. Biological hypercomputation (BH) is the concept coined that expresses that living beings process information non-algorithmically [54].

For Explainable Artificial Intelligence (XAI) there are many ways to support understanding that are not algorithmic. Explanation support can derive from aspects of the interface or architecture. Support can be provided by methods for evaluating the system and determining whether it is effective. Support can be tutorials, help documents, and even conversations among developers or users that help them understand the AI. The paper refers to these, generally, as Non-Algorithms [55].

Algorithmic security is insufficient, and many cryptographic algorithms are vulnerable to side-channel attacks (SCA), where an attacker passively or actively gets side-channel data to compromise the security properties that are assumed to be safe theoretically. This survey explores such imminent threats and their countermeasures with respect to PQC. It provides the respective, latest advancements in PQC research, as well as assessments and visions on the different types of SCAs [56].

Some notes from the remote history (year 1973) witnessed by the current author, where the notion ***algorithm*** dominated, and telling something non-algorithmic was almost a "crime", but nevertheless some bravery succeeded, like ***Non-algorithmic bit machines*** [57]. We had useful discussions with the author Prof. G.E. Pukhov of the National Academy of Sciences who was developing analogue solutions, which appeared conceptually close to self-spreading code in networks [37] (as a prototype of the current SGT).

8.6 Conclusions

The Chapter reviewed algorithms and their types, compared SGT with algorithms and their different modifications, where SGL scenarios often resemble massive spatial processes and effects like virus, radiation, smell, light, etc., and are far more general, powerful, and global than just algorithmic sequences of operations. As one of possible explanations of this universal approach, it exhibited SGT as self-evolving recursive spatial pattern capable of matching and managing large continuous spaces, and not only discrete ones, also analyzed resemblance of dynamic self-organized spatial capabilities of SGT to non-algorithmic ideas which are growing are popular especially in relation to living organisms, human societies, and self-organized systems. And finally, as the result of this Chapter, it can be stated as follows.

Traditional as well as extended algorithms are just specific tools for solving concrete problems, whereas Spatial Grasp concept represents a universal philosophy, paradigm, and technology oriented on explaining, understanding, covering, ruling and, if needed, conquering of any spatial and distributed systems, up to the whole universe.

References

1. Standard algorithms. https://en.wikipedia.org/wiki/Standard_algorithms
2. Sample standard/traditional algorithms. https://education.alberta.ca/media/481795/factsheet-algorithms.pdf
3. Algorithm. https://en.wikipedia.org/wiki/Algorithm
4. Algorithm. https://www.merriam-webster.com/dictionary/algorithm
5. Gillis, A.S.: What is an algorithm? https://www.techtarget.com/whatis/definition/algorithm
6. Algorithm. https://www.nnlm.gov/guides/data-glossary/algorithm
7. Nikolopoulou, K.: What is an algorithm? | Definition & examples. Scribbr, 9 Aug 2023. https://www.scribbr.com/ai-tools/what-is-an-algorithm/
8. What is an algorithm? HowStuffWorks, 5 Mar 2024. https://computer.howstuffworks.com/what-is-a-computer-algorithm.htm

9. Parallel algorithm. https://en.wikipedia.org/wiki/Parallel_algorithm
10. Parallel algorithms. https://www.sciencedirect.com/topics/computer-science/parallel-algorithms
11. Analysis of parallel algorithms. https://en.wikipedia.org/wiki/Analysis_of_parallel_algorithms
12. Parallel algorithm models in parallel computing, 23 July 2025. https://www.geeksforgeeks.org/mobile-computing/parallel-algorithm-models-in-parallel-computing/
13. Distributed algorithm. https://en.wikipedia.org/wiki/Distributed_algorithm
14. Distributed algorithm. ScienceDirect. Computer Science (2001). https://www.sciencedirect.com/topics/computer-science/distributed-algorithm
15. Guide to distributed algorithms. Design, analysis and implementation using python. Textbook (2025). https://doi.org/10.1007/978-3-031-79018-8
16. Mobile agent. https://en.wikipedia.org/wiki/Mobile_agent
17. Bieszczad, A., Pagurek, B., White, T.: Mobile agents for network management. IEEE Commun. Surv. **1**(1) (1998). https://doi.org/10.1109/COMST.1998.5340400
18. Silveira, R., Filho, R.S.: The mobile agents paradigm. ResearchGate, Apr 2001. https://www.researchgate.net/publication/2357161_The_Mobile_Agents_Paradigm
19. Spatial analysis. https://en.wikipedia.org/wiki/Spatial_analysis
20. Essential Geospatial Algorithms to Know for Geospatial Engineering. Fiveable Inc. (2025). https://fiveable.me/lists/essential-geospatial-algorithms
21. Nikander, J., Helminen, J.: Algorithm visualization in teaching spatial data algorithms. In: 2007 11th International Conference Information Visualization (IV '07). https://doi.org/10.1109/IV.2007.21
22. Sapaty, P.S.: A distributed processing system. European Patent N 0389655, Publ. 10.11.93. European Patent Office
23. Sapaty, P.S.: Mobile Processing in Distributed and Open Environments. Wiley, New York (1999)
24. Sapaty, P.S.: Ruling Distributed Dynamic Worlds. Wiley, New York (2005)
25. Sapaty, P.S.: Managing Distributed Dynamic Systems with Spatial Grasp Technology. Springer (2017)
26. Sapaty, P.S.: Holistic Analysis and Management of Distributed Social Systems. Springer (2018)
27. Sapaty, P.S.: Complexity in International Security: A Holistic Spatial Approach. Emerald Publishing (2019)
28. Sapaty, P.S.: Symbiosis of Real and Simulated Worlds Under Spatial Grasp Technology. Springer (2021)
29. Sapaty, P.S.: Spatial Grasp as a Model for Space-based Control and Management Systems. CRC Press (2022)
30. Sapaty, P.S.: The Spatial Grasp Model: Applications and Investigations of Distributed Dynamic Worlds. Emerald Publishing (2023)
31. Sapaty, P.S.: Providing Integrity, Awareness, and Consciousness in Distributed Dynamic Systems. CRC Press (2024)
32. Sapaty, P.S.: Spatial Networking in the United Physical, Virtual, and Mental World. Springer (2024)
33. Sapaty, P.S.: Self-Healing and Self-Recovering Systems under the Spatial Grasp Model. Emerald Publishing Limited (2025)
34. Sapaty, P.S.: Global and international security under spatial grasp paradigm. Int. Relat. Dipl. **12**(2), 72–85 (2024). https://doi.org/10.17265/2328-2134/2024.02.004
35. Sapaty, P.S.: Self-recovering-systems-under-spatial-grasp-technology. Cris. Stress. Hum. Resil. Int. J. **6**(2) (2024). https://www.crisisjournal.org/article/126490-self-recovering-systems-under-spatial-grasp-technology
36. Sapaty, P.S.: Spatial grasp model for distributed management and its comparison with traditional algorithms. Int. Relat. Dipl. **13**(3) (2025). https://doi.org/10.17265/2328-2134/2025.03.004. https://www.davidpublisher.com/Public/uploads/Contribute/689ada551780e.pdf$22

37. Sapaty, P.S.: A method of organization of an intercomputer dialogue in the radial computer systems. In: The Design of Software and Hardware for Automatic Control Systems. Institute of Cybernetics Press, Kiev (1973)
38. Virus. https://en.wikipedia.org/wiki/Virus
39. Smog. https://www.gem.wiki/Smog
40. Radiation. https://en.wikipedia.org/wiki/Radiation
41. Flood. https://en.wikipedia.org/wiki/Flood
42. Rain. https://en.wikipedia.org/wiki/Rain
43. Smell. https://en.wikipedia.org/wiki/Sense_of_smell
44. Light. https://www.wikilectures.eu/w/Light
45. Opinion. https://en.wikipedia.org/wiki/Public_opinion
46. Fear. https://mail.google.com/mail/u/0/#inbox
47. Tarancon, J.-E.P.: Non-algorithmic procedures and algorithm creation. Acta Univ. Apulensis **71** (2022). https://doi.org/10.17114/j.aua.2022.71.03
48. Lippert-Rasmussen, K.: Algorithmic and non-algorithmic fairness: should we revise our view of the latter given our view of the former? Law Philos. **44**, 155–179 (2025). https://doi.org/10.1007/s10982-024-09505-4
49. Rubin, S.: Non-algorithmic mathematical models of biological organization. Mathematics, Section E3: Mathematical biology. 30 Sept 2025. ISSN 2227-7390. https://www.mdpi.com/journal/mathematics/special_issues/Anticipation_Mathematical_Models_Biological_Causal_Organization
50. Biron, R.M.: Exploring the non-algorithmic in critical thinking. J. Thought **28**(3/4) (1993). https://www.jstor.org/stable/42589357
51. The emperor's new mind. https://en.wikipedia.org/wiki/The_Emperor's_New_Mind
52. Bringsjord, S.: Toward non-algorithmic AI. In: AI and Cognitive Science '92. Workshops in Computing. Springer, London (1993). https://doi.org/10.1007/978-1-4471-3207-3_22.
53. Zizzi, P.: The non-algorithmic side of the mind. Computer Science—Artificial Intelligence. Submitted on 15 Jan 2012. https://arxiv.org/pdf/1205.1820
54. Maldonado, C.E.: Positive affirmation of non-algorithmic information processing. Cinta moebio (60) (2017). https://doi.org/10.4067/S0717-554X2017000300279
55. Mueller, S.T., Klein, G., et al.: Non-algorithmic methods for explainable artificial intelligence. Technical report. DARPA Explainable AI Program. Applied Artificial Intelligence Letters (2021). https://data.kitware.com/api/v1/item/617af0702fa25629b9c86c02/download?contentDisposition=inline
56. Canto, A.C., Kaur, J., et al.: Algorithmic security is insufficient: a comprehensive survey on implementation attacks haunting post-quantum. Security (2023). https://doi.org/10.36227/techrxiv.23071079.v1
57. Pukhov, G.E., Evdokimov, V.F.: The sign code and especially its application in non-algorithmic bit machines. In: Pukhov, G.E. (ed.) Sbornik trudov seminara, Matematicheskoe modelirovanie i teoriya elektricheskikh tsepei, no. 11, pp. 18–20 (1973)

Chapter 9
Conclusions

Abstract This summarizes the investigation, research, analysis, and development of this book. The obtained results can be used for the development of advanced technologies for dealing with large and complex terrestrial and celestial spaces and systems, also as radically new spatially-oriented teaching and research courses and programs at universities and academic research institutions. The described Spatial Grasp Model, Technology, and Language can be easily and quickly implemented on any platform, which was investigated and prototyped for its previous versions in different countries. Multiple SGL interpreter copies integrated with other systems and advanced communications, can altogether represent powerful spatial engines effectively solving any spatial problems and operating without central resources or control.

9.1 Sources Analyzed and Used for the Book Development

These represent the following: World spatial problems [1–3]; Other books in the area [4–10]; Spatial Grasp Technology [11–27]; Ideas and activities in the spatial world [28–53]; Distributed programming languages [54–64]; Physical, virtual and hybrid spaces [65–72]; Different spatial management issues [73–80]; Spatial security and cybersecurity issues [24, 81–89]; Fighting virus sources in distributed networks [90, 91]; Removing multiple junk items in space [92]; Network security issues [93–98]; Traditional algorithms [99–106]; Parallel algorithms [107–110]; Distributed algorithms [111–113]; Mobile agents [114–116]; Spatial analysis algorithms [30, 117, 118]; Massive spatial processes [119–127]; Non-algorithmic thinking [128–138]; SGT, SGL, and WAVE history [139–195].

P. S. Sapaty, *Distributed Management with Spatial Grasp Model*, Studies in Systems, Decision and Control 666, https://doi.org/10.1007/978-3-032-19378-0_9

9.2 Spatial Aspects, Spatial Grasp Technology, and Algorithms

Spatial aspects [1–3, 24, 28–53, 65–72, 81–89] are fundamental to human activity because all actions and thought processes occur within and relate to a specific space, influencing everything from daily navigation and cognitive mapping to complex behaviors like migration, economic choices, and social organization. Spatial ability allows humans to understand, reason, and remember relationships between objects and locations, making it critical for survival, decision-making, and understanding the world, as seen in fields from human geography and urban planning to public health. The global area of such activity may symbolically look like in Fig. 9.1a.

Spatial Grasp Model and Technology* (SGT)** [11–27, 139–195] is a paradigm for managing and analyzing large-scale, distributed spaces and systems by representing tasks as self-propagating, self-modifying, active patterns written in ***Spatial Grasp Language (SGL). Instead of centralized control, SGT employs a ***self-controlled flooding*** or even ***super-virus*** approach, allowing active patterns to dynamically cover and match distributed virtual, physical, and executive (human/robotic) environments without central resources. Dynamic spatial coverage by SGT may look like in Fig. 9.1b.

Within this, a high-level operational scenario in SGL, starting in any world points propagates, ***covers and matches*** the spatial environment in ***parallel wavelike mode***. This can result in returning and analyzing the reached states and/or data, and in further waves from the initial and new points, which may be numerous and remote. Communicating SGL interpreters (potentially up to thousands and millions copies) integrated with communication systems can represent ***powerful spatial engines*** effectively operating without central resources and control.

In conclusion, it may be said that traditional and extended algorithms [30, 99–118] represent only ***specific tools*** for solving concrete problems, whereas Spatial Grasp model is a ***universal philosophy, paradigm, and technology*** oriented on explaining, covering, and ruling of the whole world.

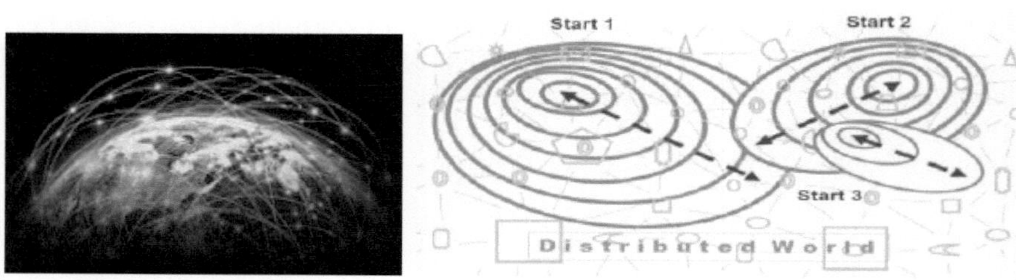

a b

Fig. 9.1 **a** Spatial area of human activity and **b** spatial grasp model covering it

9.3 Some Examples of Book Solutions

Examples of some book solutions borrowed from different chapters with related figures are summarized in Fig. 9.2, having the following brief explanations.

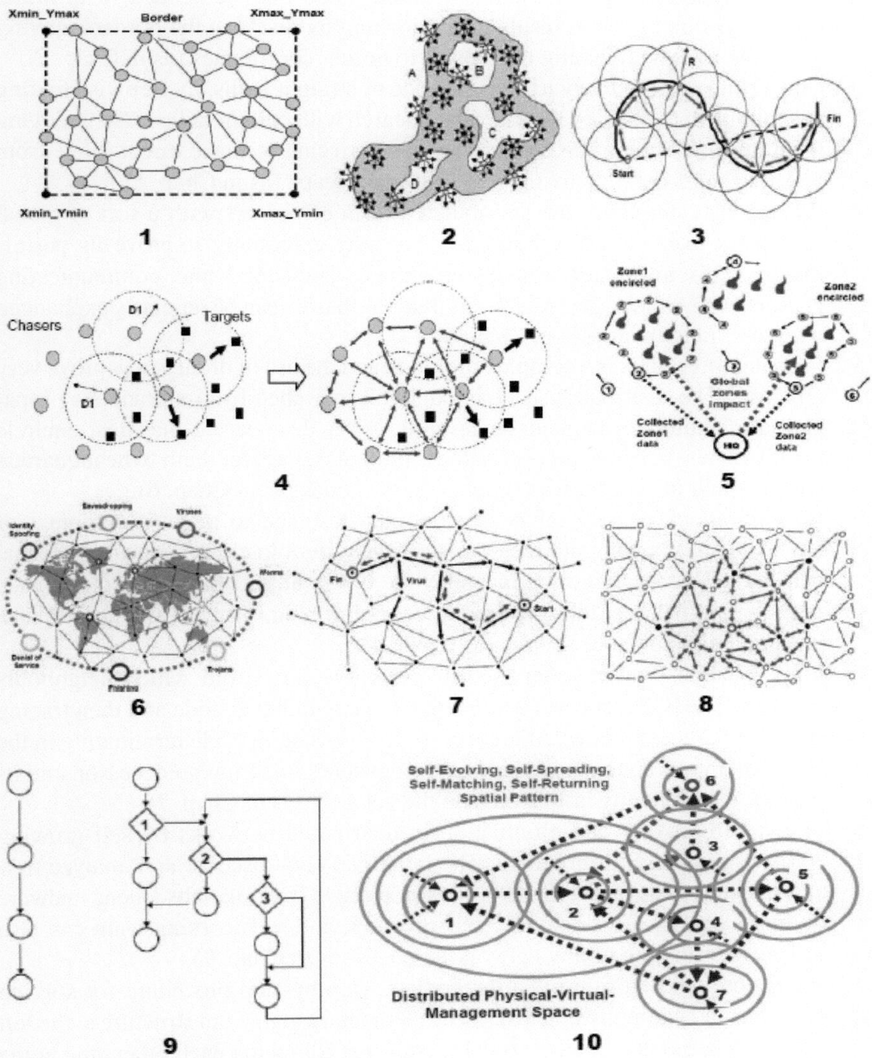

Fig. 9.2 Exemplary solutions from different chapters

1. Finding the network dimensions and surrounding borders creation. The resultant borders can also be served altogether as a security fence, if the border is supplied with proper features. Another possible solution is in finding the network area centre and creating circular border (as in Chap. 5).
2. Finding perimeter coordinates of complex spatial regions. A combination of their navigation in parallel and sequential mode may be needed, with initial massive parallel landing inside the region and by reaching the borders moving along them with collecting the closest to border coordinates (as in Chap. 5).
3. Converting a curved way to the sequence of straight paths, by stepwise finding new points in the curved road by space search with certain radius. Then moving from Start to Fin via the created paths, also trying to make direct move from Start to Fin, if possible by the territory's landscape (as in Chap. 5).
4. By regularly enriching the distributed swarm of chasers with a sort of global awareness over the operational area we may essentially improve its performance. This awareness can be effectively embedded into communicating chasers, where the targets seen by individual chasers can be regularly exchanged with their neighbours (as in Chap. 6).
5. Fighting forest fires is a complex problem, and the use of drones may prove very effective. The distributed space is randomly searched by simultaneous propagation of multiple reconnaissance units. After discovering the fires, encircle the fire zones, collect their perimeter coordinates, transfer them to headquarters (HQ), which initiate massive impact on the zones (as in Chap. 6).
6. A cyber attack usually refers to an action designed to target a computer or any element of a computerized information system to change, destroy, or steal data, as well as exploit or harm a network. Cyber attacks have been on the rise, with their types including spoofing, eavesdropping, viruses, worms, trojans, phishing, and denial of service (as in Chap. 7).
7. Finding virus sources when nodes contain records from which neighbours infected. The SGL scenario is starting from any infected node and then tracing virus source via the infected predecessors. This spatial cycle terminating in the believable virus source node having no registered infection predecessor, and its name is subsequently issued outside the network (as in Chap. 7).
8. Malware discovery and elimination in distributed networks by self-growing and self-matching virtual team. The latter can also discover and analyze new and unusual cases and effectively learn and use them in subsequent malware finding-cleaning. This spatially self-growing, and self-learning team can also self-recover in case of any network damages (as in Chap. 7).
9. At its core, an algorithm is a methodical, step-by-step procedure for solving problems or accomplishing tasks. The elementary algorithm structure at the left is reflecting the flowchart with operations just following each other, and at the right exhibiting a cyclic one with decision makers, like in finding the greatest common divisor of two numbers (as in Chap. 8).
10. SGT and SGL, both ideologically and technically, allow for a much broader and universal problem definitions and formulations, where spatial pattern-formulated tasks can be recursive, self-evolving, actually ***alive***, effectively

impacting and managing the whole continuous world. In comparison, traditional and extended algorithms may be considered only as specific tools for solving concrete problems (as in Chap. 8).

9.4 Conclusions

The described Spatial Grasp paradigm ***covers any spatial ideas*** and systems philosophically, methodologically, and practically. It ***fundamentally differs from any other models and languages*** for description, control, and management of large distributed dynamic systems traditionally representing solutions by parts (agents) communicating with each other and exchanging messages. The SGT scenarios describe all solutions on a much broader and higher level, in the form of ***integral spatial flooding, radiation, or even conscious super-virus***. These solutions can always remain alive by self-propagating, self-replicating, self-matching and self-recovering in distributed spaces. Different techniques have been described in SGL for vision, understanding and modifying of large distributed spaces, also for security, automatic control and withstanding cyber attacks in networked systems. The book finally ***compares Spatial Grasp model with algorithms*** and their modifications, confirming ***universality of the former for any spatial systems***, with the latter just representing specific tools for concrete applications.

References

1. Cantwell, J., Shukla, P.: Spatial development of technological knowledge and the evolution of international business activity across technological paradigms. Int. Bus. Rev. **34**(1) (2025). https://ideas.repec.org/a/eee/iburev/v34y2025i1s0969593124001033.html
2. Redding, S.J.: Spatial economics (2024). https://doi.org/10.3386/w33125
3. New frontiers in global security: emerging technologies and outer space. https://unu.edu/news/new-frontiers-global-security-emerging-technologies-and-outer-space
4. Eysenck, M.W., Brysbaert, M.: Fundamentals of Cognition. Routledge (2018). ISBN-10: 113867043X
5. Renz, J.: Qualitative Spatial Reasoning with Topological Information. Lecture Notes in Computer Science, 2293. Springer (2002). ISBN-10: 3540433465
6. Shkutsko, A.N., Semyachkov, K.A.: Emerging Frontiers in Spatial Development, Economic Integration, and Innovation Ecosystems: Research Anthology and Selected Contributions (2025) (Independently published). ISBN-13: 979-8311345224
7. Wieder, S., Wachs, H.: Visual/Spatial Portals to Thinking, Feeling and Movement: Advancing Competencies and Emotional Development in Children with Learning and Autism Spectrum Disorders. Profectum Foundation (2012). ISBN-10: 0578111284
8. Drewello, H., Scholl, B.: Integrated Spatial and Transport Infrastructure Development: The Case of the European North-South Corridor Rotterdam-Genoa (Contributions to Economics). Springer (2015). ISBN-10: 3319157078
9. Dixon, B.: Interdisciplinary Environmental Solutions: Using Geospatial Technologies for Bridging Disciplines, Scale and Data. Palgrave Macmillan (2025). ISBN-10: 3031167627

10. Gold, C.: Spatial Context: An Introduction to Fundamental Computer Algorithms for Spatial Analysis. ISPRS Book Series 12. CRC Press (2018). ISBN-13: 978-1498779104
11. Sapaty, P.S.: A distributed processing system. European Patent N 0389655, Publ. 10.11.93. European Patent Office
12. Sapaty, P.S.: Mobile Processing in Distributed and Open Environments. Wiley, New York (1999)
13. Sapaty, P.S.: Ruling Distributed Dynamic Worlds. Wiley, New York (2005)
14. Sapaty, P.S.: Managing Distributed Dynamic Systems with Spatial Grasp Technology. Springer (2017)
15. Sapaty, P.S.: Holistic Analysis and Management of Distributed Social Systems. Springer (2018)
16. Sapaty, P.S.: Complexity in International Security: A Holistic Spatial Approach. Emerald Publishing (2019)
17. Sapaty, P.S.: Symbiosis of Real and Simulated Worlds under Spatial Grasp Technology. Springer (2021)
18. Sapaty, P.S.: Spatial Grasp as a Model for Space-Based Control and Management Systems. CRC Press (2022)
19. Sapaty, P.S.: The Spatial Grasp Model: Applications and Investigations of Distributed Dynamic Worlds. Emerald Publishing (2023)
20. Sapaty, P.S.: Providing Integrity, Awareness, and Consciousness in Distributed Dynamic Systems. CRC Press (2024)
21. Sapaty, P.S.: Spatial Networking in the United Physical, Virtual, and Mental World. Springer (2024)
22. Sapaty, P.S.: Self-Healing and Self-Recovering Systems under the Spatial Grasp Model. Emerald Publishing Limited (2025)
23. Sapaty, P.S.: Spatial Networks as Models for Organoid Cultures and Brain Research. Mathematical Mach. Syst. (1) (2024). ISSN 1028-9763. http://www.immsp.kiev.ua/publications/articles/2024/2024_1/01_24_Sapaty.pdf
24. Sapaty, P.S.: Global and international security under spatial grasp paradigm. Int. Relat. Dipl. **12**(2), 72–85 (2024). https://doi.org/10.17265/2328-2134/2024.02.004
25. Sapaty, P.S.: Self-recovering-systems-under-spatial-grasp-technology. Cris. Stress. Hum. Resil. Int. J. **6**(2) (2024). https://www.crisisjournal.org/article/126490-self-recovering-systems-under-spatial-grasp-technology
26. Sapaty, P.S.: Spatial grasp model for distributed management and its comparison with traditional algorithms. Int. Relat. Dipl. **13**(3) (2025). https://doi.org/10.17265/2328-2134/2025.03.004. https://www.davidpublisher.com/Public/uploads/Contribute/689ada551780e.pdf
27. Sapaty, P.S.: Spatial management of distributed dynamic worlds. Cris. Stress. Hum. Resil. Int. J. **7**(1) (2025). crisisjournal.org
28. Oliver, S.: What is spatial temporal in philosophy? (2019). https://www.quora.com/What-is-spatial-temporal-in-philosophy#:~:text=This%20is%20an%20adjective%20describing,as%20your%20thoughts%20and%20emotions
29. Spacetime. https://en.wikipedia.org/wiki/Spacetime
30. Spatial analysis. https://en.wikipedia.org/wiki/Spatial_analysis
31. Spatial thinking. https://geometriedidaktik.at/en/spatial-ability/
32. Spatial vision. https://www.sciencedirect.com/topics/neuroscience/spatial-vision
33. Bardt, C.: The Feeling of Space. The MIT Press (2024). ISBN: 9780262049368. https://mitpress.mit.edu/9780262049368/the-feeling-of-space/
34. Spatial understanding. https://www.sciencedirect.com/topics/computer-science/spatial-understanding
35. Seladi-Schulman, J.: What's important about spatial awareness? 21 Apr 2020. https://www.healthline.com/health/spatial-awareness
36. Galland, D., Grønning, M.: Spatial consciousness, 15 Apr 2019. https://doi.org/10.1002/9781118568446.eurs0308
37. Spatial ability. https://en.wikipedia.org/wiki/Spatial_ability

38. Fitzgibbons, L.: Spatial intelligence. https://www.techtarget.com/whatis/definition/spatial-intelligence
39. Laurini, R., Thompson, D.: 17—Spatial knowledge: intelligent spatial information systems. In: Fundamentals of Spatial Information Systems, pp. 620–670 (1992). https://www.sciencedirect.com/science/article/abs/pii/B9780080924205500220
40. Danger space. https://en.wikipedia.org/wiki/Danger_space
41. Space warfare. https://en.wikipedia.org/wiki/Space_warfare
42. Dandoulaki, M., Lazoglou, M.: Disaster risk management and spatial planning: evidence from the fire-stricken are of Mati, Greece. Sustainability **15**(12), 9776 (2023). https://doi.org/10.3390/su15129776
43. Spatial planning. https://en.wikipedia.org/wiki/Spatial_planning
44. Nejasmic, J., Bucher, L., Knauff, M.: Grounded spatial belief revision. Acta Physiol. (Oxf.) **157**, 144–154 (2015). https://doi.org/10.1016/j.actpsy.2015.02.008
45. Corrigan, J.: Spatiality and religion. In: The Spatial Turn. Routledge (2008). https://doi.org/10.4324/9780203891308
46. Bielińska, A.: Love in the outer space. On the spatiality of being together. https://www.taylorfrancis.com/chapters/edit/10.4324/9781003436188-7/love-outer-space-agata-bieli%C5%84ska
47. Hopkins, P.: Afterword: Spatializing Hate—Relational, Intersectional and Emotional Approaches. Bristol University Press. https://doi.org/10.56687/9781529215205-017
48. Hipp, J.R.: The Spatial Scale of Crime, 1st edn. Routledge (2022). https://www.amazon.com/Spatial-Scale-Crime-John-Hipp/dp/103220236X
49. Spatial Diversity. From: Wireless Communications Design Handbook (1998). https://www.sciencedirect.com/topics/engineering/spatial-diversity
50. Funk, G.A., Jansen, V.A.A., et al.: Spatial models of virus-immune dynamics. J. Theor. Biol. **233**(2) (2005). https://www.sciencedirect.com/science/article/abs/pii/S0022519304004758
51. Space Network Security. Chapter, First Online: 01 Mar 2023. https://doi.org/10.1007/978-981-99-1125-7_5
52. Scott, L.M.: Analysis of spatial pattern. In: International Encyclopedia of the Social & Behavioral Sciences, 2nd edn., pp. 178–184. https://doi.org/10.1016/B978-0-08-097086-8.72064-2
53. Spatial art history tutorials and training. Arts & Sciences at Washington University in St. Louis. https://arthistory.wustl.edu/spatial-art-history-tutorials
54. Programming languages for distributed systems, 03 July 2024. https://www.geeksforgeeks.org/programming-languages-for-distributed-systems/
55. Java (programming language). https://en.wikipedia.org/wiki/Java_(programming_language)
56. What is Java? https://aws.amazon.com/what-is/java/
57. Python (programming language). https://en.wikipedia.org/wiki/Python_(programming_language)
58. Master Python programming. https://www.programiz.com/python-programming
59. Go (programming language). https://en.wikipedia.org/wiki/Go_(programming_language)
60. Donovan, A., Kernighan, B.: Go Programming Language. Addison-Wesley Professional Computing
61. Erlang (programming language). https://en.wikipedia.org/wiki/Erlang_(programming_language)
62. Erlang tutorial. https://www.tutorialspoint.com/erlang/index.htm
63. Scala (programming language). https://en.wikipedia.org/wiki/Scala_(programming_language)
64. Scala programming language, 11 Dec 2023. https://www.geeksforgeeks.org/scala-programming-language/
65. Karlgren, J., Kanerva, P.: High-dimensional distributed semantic spaces for utterances. Computer Science > Computation and Language, Submitted on 1 Apr 2021. https://arxiv.org/abs/2104.00424

66. Understanding distributed representations. https://deepai.org/machine-learning-glossary-and-terms/distributed-representation
67. Hermann, K.M.: Distributed representations for compositional semantics. Computer Science > Computation and Language, Submitted on 12 Nov 2014. https://arxiv.org/abs/1411.3146
68. Space representation. In: Engineering. AI Generated Definition Based on: Encyclopedia of Behavioral Neuroscience, 2nd edn. (2022). https://www.sciencedirect.com/topics/engineering/space-representation
69. Paccanaro, A., Hinton, G.E.: Learning Distributed Representations by Mapping Concepts and Relations into a Linear Space. Gatsby Computational Neuroscience Unit, UCL, London, UK. https://www.cs.utoronto.ca/~hinton/absps/icml-lre.pdf
70. Batty, M., Miller, H.J.: Representing and visualizing physical, virtual and hybrid information spaces. In: Information, Place, and Cyberspace. Advances in Spatial Science. Springer, Berlin, Heidelberg (2000). https://doi.org/10.1007/978-3-662-04027-0_8
71. Yu, H., Shaw, S.-L.: Exploring potential human activities in physical and virtual spaces. Int. J. Geogr. Inf. Sci. **22**(4), 409–430 (2008). https://doi.org/10.1080/13658810701427569
72. Wideström, J.: Connecting Physical and Virtual Spaces. Department of Computer Science and Engineering Chalmers University of Technology, Gothenburg, Sweden (2020). https://research.chalmers.se/publication/516761/file/516761_Fulltext.pdf
73. Spatial operations. ArcMap. https://desktop.arcgis.com/en/arcmap/latest/manage-data/using-sql-with-gdbs/spatial-operations.htm
74. Scholl, M., Voisard, A.: Spatial operations and map operations. In: Liu, L., Özsu, M.T. (eds.) Encyclopedia of Database Systems. Springer, Boston, MA (2009). https://doi.org/10.1007/978-0-387-39940-9_359
75. Olazabal, M., Urzelai, A., et al.: Sustainable spatial management: an integrated approach. ICE Proc. Munic. Eng. **163**(1), 33–41 (2010). https://doi.org/10.1680/muen.2010.163.1.33
76. Mamoulis, N.: Spatial Data Management. Part of the book series: Synthesis Lectures on Data Management (SLDM). Springer (2012). https://doi.org/10.1007/978-3-031-01884-8
77. Gomes, E., da Costa, E.M., Abrantes, P.: Spatial planning and land-use management. Land **13**(1), 94 (2024). https://doi.org/10.3390/land13010094
78. Carazo, M.A.: Distributed operations: the evolution of warfare in the 21st century. ResearchGate, Feb 2005. https://www.researchgate.net/publication/235158726_Distributed_Operations_The_Evolution_of_Warfare_in_the_21st_Century
79. Filipoff, D.: Executing distributed operations in an increasingly contested maritime environment. Report, 1 July 2024. https://www.atlanticcouncil.org/in-depth-research-reports/report/executing-distributed-operations-in-an-increasingly-contested-maritime-environment/
80. Sapaty, P.S.: Spatial management of air and missile defence operation. Math. Mach. Syst. (1). ISSN 1028-9763 (2023). http://www.immsp.kiev.ua/publications/articles/2023/2023_1/01_23_Sapaty.pdf
81. Yan, Y., Han, G., Xu, H.: A survey on secure routing protocols for satellite network. J. Netw. Comput. Appl. **145**(1), 102415 (2019). https://www.sciencedirect.com/science/article/abs/pii/S1084804519302498
82. Yin, Y., Huang, C., et al.: Reinforcement learning-based routing algorithm in satellite-terrestrial integrated networks. Hindawi Wirel. Commun. Mob. Comput. **2021**, Article ID 3759631. https://doi.org/10.1155/2021/3759631
83. Zhao, N., Long, X., Wang, J.: A multi-constraint optimal routing algorithm in LEO satellite networks. Wirel. Netw. (2021). https://doi.org/10.1007/s11276-021-02674-3
84. Ray, P.P.: A review on 6G for space-air-ground integrated network: key enablers, open challenges, and future direction. J. King Saud Univ. Comput. Inf. Sci. **34**(9) (2022). https://doi.org/10.1016/j.jksuci.2021.08.014
85. Zhuo, M., Liu, L., et al.: Survey on security issues of routing and anomaly detection for space information networks. Sci. Rep. https://doi.org/10.1038/s41598-021-01638-z
86. Wu, H., Sun, Y., et al.: Examining the satellite-detected urban land use spatial patterns using multidimensional fractal dimension indices. Remote Sens. **5**(10) (2013). https://doi.org/10.3390/rs5105152

87. What is cybersecurity? https://www.cisco.com/site/us/en/learn/topics/security/what-is-cybersecurity.html
88. Khan, S.K., Shiwakoti, N., et al.: Space cybersecurity challenges, mitigation techniques, anticipated readiness, and future directions. Int. J. Crit. Infrastruct. Prot. **47**, 100724 (2024). https://doi.org/10.1016/j.ijcip.2024.100724
89. Types of cyber attacks. https://www.fortinet.com/uk/resources/cyberglossary/types-of-cyber-attacks
90. Worm virus definition. https://www.fortinet.com/uk/resources/cyberglossary/worm-virus#:~:text=A%20worm%20virus%20refers%20to,virus%20needs%20a%20host%20computer
91. How can you remove malware from a network without disruption? Last updated on 29 May 2025. https://www.linkedin.com/advice/0/how-can-you-remove-malware-from-network-without-dwj2c
92. Iyer, V.: How do you clean up 170 million pieces of space junk? 05.24.23. https://fas.org/publication/how-do-you-clean-up-170-million-pieces-of-space-junk/
93. Network security. https://en.wikipedia.org/wiki/Network_security#cite_note-1
94. Network infrastructure security—best practices and strategies. https://www.dataguard.com/blog/network-infrastructure-security-best-practices-and-strategies/
95. Peterson, S., Hilliard, J.: What is a network disaster recovery plan? TechTarget. Published: 14 Dec 2022 (2022). https://www.techtarget.com/searchdisasterrecovery/definition/Network-disaster-recovery-plan
96. Liu, J., Bai, L., Jiang, C., Zhang, W.: Space network security. In: Space-Air-Ground Integrated Network Security. Springer, Singapore (2023). https://doi.org/10.1007/978-981-99-1125-7_5
97. Wu, X., Du, Y., Fan, T., et al.: Threat analysis for space information network based on network security attributes: a review. Complex Intell. Syst. **9**, 3429–3468 (2023). https://doi.org/10.1007/s40747-022-00899-z
98. Smurf attack. https://en.wikipedia.org/wiki/Smurf_attack
99. Standard algorithms. https://en.wikipedia.org/wiki/Standard_algorithms
100. Sample standard/traditional algorithms. https://education.alberta.ca/media/481795/factsheet-algorithms.pdf
101. Algorithm. https://en.wikipedia.org/wiki/Algorithm
102. Algorithm. https://www.merriam-webster.com/dictionary/algorithm
103. Gillis, A.S.: What is an algorithm? https://www.techtarget.com/whatis/definition/algorithm
104. Algorithm. https://www.nnlm.gov/guides/data-glossary/algorithm
105. Nikolopoulou, K.: What is an algorithm? | Definition & examples. Scribbr, 9 Aug 2023. https://www.scribbr.com/ai-tools/what-is-an-algorithm/
106. What is an algorithm? HowStuffWorks, 5 Mar 2024. https://computer.howstuffworks.com/what-is-a-computer-algorithm.htm
107. Parallel algorithm. https://en.wikipedia.org/wiki/Parallel_algorithm
108. Parallel algorithms. https://www.sciencedirect.com/topics/computer-science/parallel-algorithms
109. Analysis of parallel algorithms. https://en.wikipedia.org/wiki/Analysis_of_parallel_algorithms
110. Parallel algorithm models in parallel computing: 23 July 2025. https://www.geeksforgeeks.org/mobile-computing/parallel-algorithm-models-in-parallel-computing/
111. Distributed algorithm. https://en.wikipedia.org/wiki/Distributed_algorithm
112. Distributed algorithm. ScienceDirect, Computer Science (2001). https://www.sciencedirect.com/topics/computer-science/distributed-algorithm
113. Guide to distributed algorithms. In: Design, Analysis and Implementation Using Python. Textbook (2025). https://doi.org/10.1007/978-3-031-79018-8
114. Mobile agent. https://en.wikipedia.org/wiki/Mobile_agent
115. Bieszczad, A., Pagurek, B., White, T.: Mobile agents for network management. IEEE Commun. Surv. **1**(1) (1998). https://doi.org/10.1109/COMST.1998.5340400

116. Silveira, R., Filho, R.S.: The mobile agents paradigm. ResearchGate, Apr 2001. https://www.researchgate.net/publication/2357161_The_Mobile_Agents_Paradigm
117. Essential Geospatial Algorithms to Know for Geospatial Engineering. Fiveable Inc. (2025). https://fiveable.me/lists/essential-geospatial-algorithms
118. Nikander, J., Helminen, J.: Algorithm visualization in teaching spatial data algorithms. In: 2007 11th International Conference Information Visualization (IV '07). https://doi.org/10.1109/IV.2007.21
119. Virus. https://en.wikipedia.org/wiki/Virus
120. Smog. https://www.gem.wiki/Smog
121. Radiation. https://en.wikipedia.org/wiki/Radiation
122. Flood. https://en.wikipedia.org/wiki/Flood
123. Rain. https://en.wikipedia.org/wiki/Rain
124. Smell. https://en.wikipedia.org/wiki/Sense_of_smell
125. Light. https://www.wikilectures.eu/w/Light
126. Opinion. https://en.wikipedia.org/wiki/Public_opinion
127. Fear. https://mail.google.com/mail/u/0/#inbox
128. Tarancon, J.-E.P.: Non-algorithmic procedures and algorithm creation. Acta Univ. Apulensis **71** (2022). https://doi.org/10.17114/j.aua.2022.71.03
129. Lippert-Rasmussen, K.: Algorithmic and non-algorithmic fairness: should we revise our view of the latter given our view of the former? Law Philos. **44**, 155–179 (2025). https://doi.org/10.1007/s10982-024-09505-4
130. Rubin, S.: Non-algorithmic mathematical models of biological organization. Mathematics. Section E3: Mathematical Biology. 30 Sept 2025. ISSN 2227-7390. https://www.mdpi.com/journal/mathematics/special_issues/Anticipation_Mathematical_Models_Biological_Causal_Organization
131. Biron, R.M.: Exploring the non-algorithmic in critical thinking. J. Thought **28**(3/4) (1993). https://www.jstor.org/stable/42589357
132. The emperor's new mind. https://en.wikipedia.org/wiki/The_Emperor's_New_Mind
133. Bringsjord, S.: Toward non-algorithmic AI. In: AI and Cognitive Science '92. Workshops in Computing. Springer, London (1993). https://doi.org/10.1007/978-1-4471-3207-3_22
134. Zizzi, P.: The non-algorithmic side of the mind. Computer Science—Artificial Intelligence. Submitted on 15 Jan 2012. https://arxiv.org/pdf/1205.1820
135. Maldonado, C.E.: Positive affirmation of non-algorithmic information processing. Cinta moebio (60) (2017). https://doi.org/10.4067/S0717-554X2017000300279
136. Mueller, S.T., Klein, G., et al.: Non-algorithmic methods for explainable artificial intelligence. Technical report, DARPA Explainable AI Program. Applied Artificial Intelligence Letters (2021). https://data.kitware.com/api/v1/item/617af0702fa25629b9c86c02/download?contentDisposition=inline
137. Canto, A.C., Kaur, J., et al.: Algorithmic security is insufficient: a comprehensive survey on implementation attacks haunting post-quantum. Security (2023). https://doi.org/10.36227/techrxiv.23071079.v1
138. Pukhov, G.E., Evdokimov, V.F.: The sign code and especially its application in non-algorithmic bit machines. In: Pukhov, G.E. (ed.) Sbornik trudov seminara, Matematicheskoe modelirovanie i teoriya elektricheskikh tsepei, no. 11, pp. 18–20 (1973)
139. Bondarenko, A.T., Mikhalevich, S.B., Nikitin, A.I., Sapaty, P.S.: Software of BESM-6 computer for communication with peripheral computers via telephone channels. In: Computer Software, vol. 5. Institute of Cybernetics Press, Kiev (1970)
140. Sapaty, P.S.: A method of organization of an intercomputer dialogue in the radial computer systems. In: The Design of Software and Hardware for Automatic Control Systems. Institute of Cybernetics Press, Kiev (1973)
141. Sapaty, S.: A wave language for parallel processing of semantic networks. Comput. Artif. Intell. **5**(4), 289–314 (1986)
142. Sapaty, P.S.: Distributed modeling of cooperative behavior by mobile agents. In: Proceedings of Sixth Conference on Computer Generated Forces and Behavioral Representation, IST UCF, Orlando, FL, July 1996, pp. 599–613

143. Sapaty, P.S.: Distributed artificial brain for collectively behaving mobile robots. In: Proceedings of Symposium and Exhibition Unmanned Systems 2001, Baltimore, MD, July 31–Aug 2, 2001
144. Sapaty, P.S.: Crisis management with distributed processing technology. Int. Trans. Syst. Sci. Appl. **1**(1), 81–92 (2006). ISSN 1751-1461
145. Sapaty, P.S.: Distributed technology for global dominance. In: Proceedings of International Conference Defense Transformation and Net-Centric Systems 2008, as part of the SPIE Defense and Security Symposium. World Center Marriott Resort and Convention Center, Orlando, FL, USA, 16–20 Mar 2008 (Also: Proceedings of SPIE, vol. 6981. In: Suresh R (ed.) Defense Transformation and Net-Centric Systems 2008, 69810T, 3 Apr 2008)
146. Sapaty, P.S.: Distributed technology for global dominance. Keynote lecture. In: Proceedings of the Fifth International Conference in Control, Automation and Robotics ICINCO 2008, The Conference Proceedings, Funchal, Madeira, Portugal, 11–15 May 2008
147. Sapaty, P.S.: Human-robotic teaming: a compromised solution. In: AUVSI's Unmanned Systems North America 2008, San Diego, USA, 10–12 June 2008
148. Sapaty, P.S.: High-level communication protocol for dynamically networked battlefields. In: Proceedings of International Conference Tactical Communications 2009 (Situational Awareness and Operational Effectiveness in the Last Tactical Mile), One Whitehall Place, Whitehall Suite and Reception, London, UK (2009)
149. Sapaty, P.S.: Distributed capability for battlespace dominance. In: Electronic Warfare 2009 Conference and Exhibition. Novotel London West Hotel and Conference Center, London, 14–15 May 2009
150. Sapaty, P.S.: Providing spatial integrity for distributed unmanned systems. In: Proceedings of 6th International Conference in Control, Automation and Robotics ICINCO 2009, Milan, Italy (2009)
151. Sapaty, P.S.: Distributed technology for global control. Book chapter, Lecture Notes in Electrical Engineering, Informatics in Control, Automation and Robotics, vol. 37. Springer, Berlin (2009)
152. Sapaty, P.S.: Gestalt-based integrity of distributed networked systems. In: SPIE Europe Security + Defence. bcc Berliner Congress Centre, Berlin, Germany (2009)
153. Sapaty, P.S.: Remote control of open groups of remote sensors. In: Proceedings of SPIE Europe Security + Defence, Berlin, Germany (2009)
154. Sapaty, P.S.: Tactical communications in advanced systems for asymmetric operations. In: Proceedings of Tactical Communications 2010, CCT Venues, Canary Wharf, London, UK, 28–30 Apr 2010
155. Sapaty, P.S.: High-level technology to manage distributed robotized systems. In: Proceedings of Military Robotics 2010, London, UK, 25–27 May 2010
156. Sapaty, P.S.: Emerging asymmetric threats, Q&A session. In: Tactical Communications 2010, CCT Venues, Canary Wharf, London, UK, 28–30 Apr 2010
157. Sapaty, P.S.: High-level organization and management of directed energy systems. In: Proceedings of Directed Energy Weapons 2010, CCT, Canary Wharf, London, UK, 25–26 Mar 2010
158. Sapaty, P.S.: Unified transition to robotized armies with spatial grasp technology. In: International Summit Military Robotics, London, United Kingdom, 12–13 Nov 2012
159. Sapaty, P.S.: Distributed air and missile defense with spatial grasp technology. Intell. Control Autom. Sci. Res. **3**(2), 117–131 (2012)
160. Sapaty, P.S.: Global electronic dominance. In: 12th International Fighter Symposium, 6th–8th Nov 2012, Grand Connaught Rooms, London, UK
161. Sapaty, P.S.: Providing global awareness in distributed dynamic environments. In: International Summit ISR, London, 16–18 Apr 2013
162. Sapaty, P.S.: Ruling distributed dynamic worlds with spatial grasp technology. Tutorial at the International Science and Information Conference 2013 (SAI), London, UK, 7–9 Oct 2013
163. Sapaty, P.S.: Night vision under advanced spatial intelligence: a key to battlefield dominance. In: International Summit Night Vision 2013, London, 4–6 June 2013

164. Sapaty, P.S.: Integration of ISR with advanced command and control for critical mission applications. In: SMi's ISR Conference, Holiday Inn Regents Park, London, 7–8 Apr 2014
165. Sapaty, P.S.: Unified transition to cooperative unmanned systems under spatial grasp paradigm. Int. J. Trans. Netw. Commun. **2**(2) (2014)
166. Sapaty, P.S.: From manned to smart unmanned systems: a unified transition. In: SMi's Military Robotics, Holiday Inn Regents Park, London, 21–22 May 2014
167. Sapaty, P.S.: Unified transition to cooperative unmanned systems under spatial grasp paradigm. In: 19th International Command and Control Research and Technology Symposium, Alexandria, Virginia, 16–19 June 2014
168. Sapaty, P.S.: Distributed human terrain operations for solving national and international problems. Int. Relat. Dipl. **2**(9) (2014)
169. Sapaty, P.S.: Providing over-operability of advanced ISR systems by a high-level networking technology. In: SMI's Airborne ISR, Holiday Inn Kensington Forum, London, United Kingdom, 26–27 Oct 2015
170. Sapaty, P.S.: Distributed missile defence with spatial grasp technology. In: SMi's Military Space, Holiday Inn Regents Park, London, 4–5 Mar 2015
171. Sapaty, P.S.: Military robotics: latest trends and spatial grasp solutions. Int. J. Adv. Res. Artif. Intell. **4**(4) (2015)
172. Sapaty, P.S.: A brief introduction to the spatial grasp language (SGL). J. Comput. Sci. Syst. Biol. **09**(02) (2015). https://www.researchgate.net/publication/307744279_A_Brief_Introduction_to_the_Spatial_Grasp_Language_SGL
173. Sapaty, P.S.: Organization of advanced ISR systems by high-level networking technology. Math. Mach. Syst. (1) (2016). http://www.immsp.kiev.ua/publications/articles/2016/2016_1/01_2016_Sapaty.pdf
174. Sapaty, P.S.: Towards massively robotized systems under spatial grasp technology. J. Comput. Sci. Syst. Biol. **9**(1) (2016)
175. Sapaty, P.S.: Towards wholeness and integrity of distributed dynamic systems. J. Comput. Sci. Syst. Biol. **9**(3) (2016)
176. Sapaty, P.S.: Towards global goal orientation, robustness and integrity of distributed dynamic systems. J. Int. Relat. Dipl. **4**(6) (2016)
177. Sapaty, P.: Spatial grasp language (SGL). In: Advances in Image and Video Processing, vol. 4, no. 1 (2016). https://journals.scholarpublishing.org/index.php/AIVP/article/view/1922/1052
178. Sapaty, P.S.: Mosaic warfare: from philosophy to model to solution. Mathematical Mach. Syst. (3) (2019)
179. Sapaty, P.S.: Global network management under spatial grasp paradigm. Int. Robot. Autom. J. **6**(3), 134–148 (2020). https://medcraveonline.com/IRATJ/IRATJ-06-00212.pdf
180. Sapaty, P.S.: Global network management under spatial grasp paradigm. Glob. J. Res. Eng. J. Gen. Eng. **20**(5), 58–81. Version 1.0 (2020). https://globaljournals.org/GJRE_Volume20/6-Global-Network-Management.pdf
181. Sapaty, P.: Advanced terrestrial and celestial missions under spatial grasp technology. Aeronaut. Aerosp. Open Access J. **4**(3) (2020). https://medcraveonline.com/AAOAJ/AAOAJ-04-00110.pdf
182. Sapaty, P.: Spatial management of distributed social systems. J. Comput. Sci. Res. **02**(03) (2020). https://ojs.bilpublishing.com/index.php/jcsr/article/view/2077/pdf
183. Sapaty, P.: Towards global nanosystems under high-level networking technology. Acta Sci. Comput. Sci. **2**(8) (2020). https://www.actascientific.com/ASCS/pdf/ASCS-02-0051.pdf
184. Sapaty, P.: Symbiosis of distributed simulation and control under spatial grasp technology. SSRG Int. J. Mob. Comput. Appl. (IJMCA) **7**(2) (2020). http://www.internationaljournalssrg.org/IJMCA/2020/Volume7-Issue2/IJMCA-V7I2P101.pdf
185. Sapaty, P.S.: Symbiosis of real and simulated worlds under global awareness and consciousness. In: Abstract, The Science of Consciousness (TSC) Conferences, Tucson, Arizona (2020). https://eagle.sbs.arizona.edu/sc/report_poster_detail.php?abs=3696
186. Sapaty, P.S.: Spatial grasp as a model for space-based control and management systems. Math. Mach. Syst. (1), 135–138 (2021). http://www.immsp.kiev.ua/publications/articles/2021/2021_1/Sapaty_book_1_2021.pdf

187. Sapaty, P.S.: Managing multiple satellite architectures by spatial grasp technology. Math. Mach. Syst. (1), 3–16 (2021). http://www.immsp.kiev.ua/publications/eng/2021_1/
188. Sapaty, P.S.: Spatial management of large constellations of small satellites. Math. Mach. Syst. (2) (2021). http://www.immsp.kiev.ua/publications/articles/2021/2021_2/02_21_Sapaty.pdf
189. Sapaty, P.S.: Global management of space debris removal under spatial grasp technology. Acta Sci. Comput. Sci. **3**(7) (2021). https://www.actascientific.com/ASCS/pdf/ASCS-03-0135.pdf
190. Sapaty, P.S.: Space debris removal under spatial grasp technology. Netw. Commun. Technol. **6**(1) (2021). https://www.ccsenet.org/journal/index.php/nct/article/view/0/45486
191. Sapaty, P.S.: Relation of spatial grasp paradigm to higher psychological and mental concepts. Acta Sci. Comput. Sci. **4**(12) (2022). https://actascientific.com/ASCS/pdf/ASCS-04-0359.pdf
192. Sapaty, P.S.: Seeing and managing distributed worlds with spatial grasp paradigm. Acta Sci. Comput. Sci. **4**(12) (2022). https://actascientific.com/ASCS/pdf/ASCS-04-0365.pdf
193. Sapaty, P.S.: Comprehending distributed worlds with the spatial grasp paradigm. Math. Mach. Syst. (1) (2022). http://www.immsp.kiev.ua/publications/articles/2022/2022_1/01_22_Sapaty.pdf
194. Sapaty, P.S.: Providing global awareness in distributed dynamic systems. Int. Relat. Dipl. **11**(2), 87–100 (2023). https://doi.org/10.17265/2328-2134/2023.02.002. https://www.davidpublisher.com/Public/uploads/Contribute/6486c3d05a6cc.pdf
195. Sapaty, P.S.: Simulating distributed consciousness with spatial grasp model. Math. Mach. Syst. (3) (2023). http://www.immsp.kiev.ua/publications/articles/2023/2023_3/03_23_Sapaty.pdf

MIX
Papier aus verantwortungsvollen Quellen
Paper from responsible sources
FSC® C105338

If you have any concerns about our products,
you can contact us on
ProductSafety@springernature.com

In case Publisher is established outside the EU,
the EU authorized representative is:
Springer Nature Customer Service Center GmbH
Europaplatz 3, 69115 Heidelberg, Germany

Printed by Libri Plureos GmbH
in Hamburg, Germany